Magnifying Creation

George T. Javor, PhD

TEACH Services, Inc.
P U B L I S H I N G
www.TEACHServices.com • (800) 367-1844

World rights reserved. This book or any portion thereof may not be copied or reproduced in any form or manner whatever, except as provided by law, without the written permission of the publisher, except by a reviewer who may quote brief passages in a review.

The author assumes full responsibility for the accuracy of all facts and quotations as cited in this book. The opinions expressed in this book are the author's personal views and interpretations, and do not necessarily reflect those of the publisher.

This book is provided with the understanding that the publisher is not engaged in giving spiritual, legal, medical, or other professional advice. If authoritative advice is needed, the reader should seek the counsel of a competent professional.

Copyright © 2024 George T. Javor, PhD
Copyright © 2024 TEACH Services, Inc.
ISBN-13: 978-1-4796-1750-0 (Paperback)
ISBN-13: 978-1-4796-1751-7 (ePub)
Library of Congress Control Number: 2024916457

Unless otherwise indicated, all scriptures in this book are from the King James Version.

Scriptures credited to NCV are quoted from The Holy Bible, New Century Version, copyright © 2005 by Thomas Nelson, Inc. Used by permission.

Scriptures credited to NIV are quoted from The Holy Bible New International Version®. Copyright © 1973, 1978, 1984 by International Bible Society. All rights reserved.

Scriptures credited to NKJV are quoted from New King James Version®, copyright © 1982 by Thomas Nelson. Used by permission. All rights reserved.

Scriptures credited to NLT are quoted from Holy Bible, New Living Translation, copyright © 1996, 2004, 2015 by Tyndale House Foundation. Used by permission of Tyndale House Publishers, Inc., Carol Stream, Illinois 60188. All rights reserved.

Published by

TEACH Services, Inc.
P U B L I S H I N G
www.TEACHServices.com • (800) 367-1844

Contents

1. Magnifying Creation 9
2. The Apollo 16 Mission and Biochemical Evolution 15
3. The Billion Dollar Question 20
4. A Creationist's View of the Solar System 27
5. The Slum of the Universe 33
6. Biblical Approaches to Biology 39
7. Teaching Biology in the Light of Creation 51
8. Searching for the Creator Through a Study of a Bacterium 59
9. The Mystery of Life 71
10. Creation in Focus 81
11. It is a Wonderful Life! 89
12. Materialistic or Superintended Creation? 93
13. Proving Creation 97
14. The Scandal of Biochemical Evolution 101
15. Consequences of Creationism 106
Appendix: The Non-equilibrium State of Living Matter
Is a Significant Barrier to Laboratory Abiogenesis 113

Note: This book combines new articles with chapters from my previous book, *Evidences for Creation* (chapters 2, 3, 6, 8, and 15 above), which is accessible online at https://1ref.us/gjec although it is now out of print in book form.

Dedication

This book is dedicated to all those who choose to live in the light of creation.

Preface

Dear Reader,

This collection of essays invites you to consider topics of everlasting significance. Examining aspects of the great subject of creation draws us closer to our Savior and Best Friend.

Life without the light of creation is confusing at the minimum because we have to reconcile the great complexity of existence with its ultimate meaninglessness.

Creation, in turn, gives meaning to existence, both past and future.

Should you wish to discuss, comment, or question anything in these chapters, your feedback would be greatly appreciated. You may send it to javorg0@gmail.com or https://1ref.us/emjavor.

May the Good Lord bless you!

CHAPTER 1

Magnifying Creation

In today's evolution-obsessed world, "creation" (normally an eight-letter word) is treated as a four-letter word, twice over. There is deep irony in mainstream science's disdain of creationism, as the founders of modern science—the likes of Isaac Newton, Louis Pasteur, Michael Faraday, Robert Boyle, Francis Bacon, and many others—were God-fearing creationists. They all turned to science to discover how the created world works.

Two centuries of diligent work by thousands of dedicated scientists have yielded a very impressive body of knowledge regarding physics, chemistry, and biology. Applications of these advances gave us electricity, the computer, the automobile, the airplane, nuclear energy, antibiotics, the complete nucleotide sequence of the human genome, and the cloning of Dolly the sheep. Scientists have assumed the enviable position in society of being the priests at the altar of knowledge.

If the sum of human knowledge could be represented by the volume of a large, continuously expanding sphere, with the space outside this sphere being the "great unknown," then the surface of this sphere, as the interface between the "known" and the "unknown," would represent the extent of our ignorance. Accordingly, as the sphere of knowledge increases, so does the extent of our ignorance. It would be expected that this increase in our collective ignorance would result in a corresponding increase in our humility.

But, alas, the opposite is occurring—we are witnessing human hubris at an all-time high. Thought leaders now proclaim that the universe exists because of accidental, miraculous events. According to them, the universe originated with a "Big Bang" (first there was nothing and then it exploded) and our planet coalesced out of the "solar nebula."

It does not bother them that they do not have a clue as to how life originated on earth. The best they can do is to suggest that life emerged from either a "primordial soup" or in deep-sea thermal vents. They are not brought up short by the mere fact that they are unable to create any form of living matter and are unable to explain the very nature of matter, gravity, or magnetism. Neither can they explain where order and beauty come from.

At the apex of this sorry attitude is the universal denial of creation. If we are here as the result of an endless series of accidents, there is no reason to even consider meaning, purpose, or prospects of existence.

> It is one thing to acknowledge the biblical account of our origins as true, and it is another to live in the light of the creation.

Against this dark background shine the brilliant beams of light of the first words of Genesis: "In the beginning God created the heavens [the sky] and the earth." These are our true origins, the reason for our existence and our destiny.

The first chapter of Genesis is an eyewitness account of the creation. How authentic is it? The prophetic pen of Ellen White wrote: "Adam had learned from the Creator the history of creation; he himself witnessed the events of nine centuries; and he imparted his knowledge to his descendants" (*Patriarchs and Prophets*, p. 83). The offsprings of Shem preserved the story of the creation and carefully passed it down to Moses.

The biblical story of the creation in the first chapter of Genesis is unique among dozens of creation stories of many cultures in that it matches precisely the spoken words of the Creator (see Gen. 2:1–3; Exod. 20:8–11).

It is one thing to acknowledge the biblical account of our origins as true, and it is another to live in the light of the creation. With apologies for adapting the words of the Apostle James, if you believe in the biblical account of the creation, you do well. The devils also know that it is the LORD who created the earth—and they tremble.

The creation account of Genesis is meant to impact our lives. First, we ought to be aware continually that we are here because the greatest Mind of the universe willed it so. His benevolent genius, which is beyond human comprehension, made ample provisions for our present and future well-being and happiness.

Second, the creation account is clear that all humanity descended from Adam and Eve. Of the nearly seven billion persons inhabiting planet earth, every one of them is our blood relative! This important concept should govern our interactions with our fellow humans. Recent analysis of the complete human genome confirms humanity's oneness. Our genetic material is so similar between us all that there are no specific genes that can be used to determine a person's race.

The Creator placed the solar system into a universe already teaming with worlds populated with created beings (Job 38:4-7). Our arrival was greeted with great joy; we live in a friendly universe! (This should be news to Hollywood, which thrives on scaring movie audiences by the prospects of alien invasions of earth. Additionally, the late mathematical genius Stephen Hawking suggested that beings from other planets would be envious of our resources, and he warned against trusting visitors from outer space should they ever materialize.)

The Creator does not waste His resources; there is always deep significance in all His actions. Therefore, it is a legitimate question to ask: Why were we created? What benefit could we be to this almost limitless universe?

"All heaven took a deep and joyful interest in the creation of the world and of man. Human beings were a new and distinct order. They were made 'in the image of God,' and it was the Creator's design that they should populate the earth," wrote Ellen White in the February 11, 1902, issue of the *Review and Herald*.

It is certain that our world was meant to be an integral part of the community of created beings throughout the universe. Humans bearing the image of the Creator, both in outward resemblance and in character, were well suited to become special ambassadors of God to other created beings, assisting the LORD to forge new links with His creatures. Humanity was created to bless the universe.

Since being made in the "image of God" is what distinguishes us from all other created beings, it is worth considering just what human qualities these are. The natural place to look for such qualities is in the record of the thirty-three years the Creator lived among us. The Gospels describe Jesus as a servant-leader, healing the sick, teaching the multitudes, earnestly caring for the poor and the disadvantaged. He was creating a better world around Himself, and, even when He rebuked someone, it was an act of love.

Likewise, we are to seek a nobility of character, striving for excellence in all our endeavors, unselfishly caring for those within our sphere of influence. However, it would be surprising if beings on planets elsewhere in the universe would not also exhibit similar qualities. Therefore it may be that the uniqueness of humans among all other created beings does not consist of unique qualities or capabilities but, rather, of the extent to which our good qualities may be developed.

Speaking of humans, who after the Flood set out to build the tower of Babel, the LORD said: "Behold, the people is one, and they have all one language; and this they begin to do: and now nothing will be restrained from them, which they have imagined to do" (Gen. 11:6). The LORD acknowledges their almost unlimited potential. Under the guidance of the Creator, our potential is limitless!

There may exist a universal language spoken throughout the universe, which was also spoken by Adam, Eve, and the antediluvians for more than 1500 years. While everything on our newly created earth was likely to be unique in the universe, as our LORD is not a cookie-cutter Creator, everything was also likely to be compatible with what is found in other worlds. Visitors from other worlds would have likely been able to breathe our air and enjoy eating our unique array of fruits, vegetables, and other produce.

Adam and Eve's rebellion changed all this. We now feel isolated on earth, being the only planet in the solar system with life and finding ourselves 4.37 light-years (over 25 trillion miles) from Alpha Centauri, the nearest star system. For decades astronomers have spent millions of dollars vainly scanning the cosmos for intelligent signals.

The irony, as mentioned above, is that the Creator became a human being and lived among us for three decades. This incredible fact is apparently lost on today's leading scientific thought leaders. If current trends hold up, they will continue their spectacularly fruitless efforts to contact extraterrestrials till the end of time.

Meanwhile, believers of the biblical narrative, without the benefit of sophisticated radio telescope arrays, are successfully communicating with the Creator through prayer and are proclaiming the soon, visible return of the Creator to earth.

The doctrine of the creation is foundational to all other teachings of the Judeo-Christian faith. The first words of the book of Genesis serve as the gateway to the entire content of the Holy Book. A rejection of the literal reading of the creation story renders the entire message of the Bible inaccessible.

This becomes apparent when we consider the three main themes of the Bible: the creation of a perfect world, the fall of mankind because of the disobedience of Adam and Eve, and the restoration of mankind through redemption by Jesus Christ.

The Creator's solicitude for His creation did not cease at the end of the creation week. Although modern science treats all natural phenomena as manifestations of the automatic operation of nature, in reality, the LORD did not create our world as a sophisticated machine, operating independent of the Creator.

"It is not to be supposed that a law is set in motion for the seed to work itself, that the leaf appears because it must do so of itself. God has laws that He has instituted, but they are only the servants through which He effects results. It is through the immediate agency of God that every tiny seed breaks through the earth, and springs into life. Every leaf grows, every flower blooms, by the power of God" (Ellen G. White, *Selected Messages*, bk. 1, p. 294).

Creationists usually talk about the past exploits of the LORD, sometimes overlooking that without the Creator upholding everything, nanosecond by nanosecond, everything would cease to exist (Heb. 1:1–3). It may be that all the laws of physics, chemistry, and biochemistry, which govern the behavior of matter, are dependent on the continual expression of the Creator's power. This power provides for the continual existence of subatomic particles and for the phenomena of gravity, magnetic forces, and positive and negative charges.

The alternative is the concept that the created world exists independent of its Creator. This position is at the heart of materialism, the driving philosophy of evolution.

On the weekly Sabbath, when we remember the Great Creator for His amazing work of bringing this world into being, we ought also to thank Him for our continued existence. Thus, Sabbath observance becomes not only a celebration of the past but also of the present!

"O magnify the LORD with me, and let us exalt his name together" (Ps. 34:3).

"By the word of the LORD were the heavens made; and all the host of them by the breath of his mouth" (Ps. 33:6).

"I will praise thee, O LORD, with my whole heart; I will shew forth all the marvellous works" (Ps. 9:1).

CHAPTER 2

The Apollo 16 Mission and Biochemical Evolution
(WITH GERALD E. SNOW)

During the Apollo 16 mission between April 21 and 23, 1972, astronaut John Young exposed about 200 frames of film in a special camera placed on the surface of the moon. At times the camera was aimed at Earth and its upper atmosphere, and pictures were taken in the far ultraviolet region. The purpose was to study the gaseous contents of our upper atmosphere.

After the astronauts returned, the film in the camera was developed and analyzed. The initial observations were communicated soon afterward by a news release from the Naval Research Laboratory in Washington, D. C. It stated in part: "Solar effects on the earth's water that evaporates to the high atmosphere may provide our primary supply of oxygen, and not photosynthesis as is generally believed" (Washington, D.C.: Naval Research Laboratory News Release 30-72-7; AP, *New York Times*, July 12, 1972, p. 24).

This bit of news may not sound unusually exciting. After all, so long as we have oxygen to breathe, what is the difference whether it comes directly from water vapor in the upper atmosphere (a simple chemical process) or from photosynthesis in living, green plants? But if substantial amounts of oxygen could be produced in the absence of photosynthesis, it could have implications concerning the various theories of biochemical evolution that have been proposed to this day.

A cornerstone of modern evolutionary thought is that the first living cell came into existence from lifeless, quite simple substances found in a postulated "primitive atmosphere." This process, called "biochemical evolution," deals with a hypothetical series of events that could convert inanimate matter into living cells. Biochemical evolution logically serves as the foundation upon which the theories of biological evolution are built.

A number of laboratories throughout the world are engaged in research attempting to discover means by which components of living cells could be produced from mixtures of gases such as carbon dioxide, carbon monoxide, ammonia, methane, hydrogen sulfide, nitrogen, hydrogen, and water vapor (Sidney W. Fox, Kaoru Harada, Gottfried Krampitz, George Mueller, "Chemical Origins of Cells," *Chemical and Engineering News*, Vol. 48, Iss. 26 [22 June 1970], pp. 80–94). Various mixtures of these gases have been used, but all investigators agree in excluding oxygen. They assume that various forces in nature, such as heat, electric discharges, ultraviolet radiation from the sun, cosmic rays, and natural radioactivity, caused the substances of the primitive atmosphere to combine into biologically active compounds in the absence of oxygen (Richard M. Lemmon, "Chemical Evolution," *Chemical Reviews*, Vol. 70, Iss. 1 [1 Feb. 1970], pp. 95–109).

Two of the most important classes of chemical compounds in living cells are proteins and nucleic acids. Evolutionary theorists feel that if they can establish the processes by which these two all-important classes of organic compounds may be produced by the forces of nature alone, they have succeeded in proving the feasibility of biochemical evolution. Proteins and nucleic acids are each composed of large numbers of much smaller units. The smaller units making up proteins are amino acids, and the units that make up nucleic acids are nucleotides. There are twenty kinds of amino acids and five kinds of nucleotides. The specific order of these small units of amino acids and nucleotides is highly important. Just as, in written languages, letters must occur in a specified sequence to spell a meaningful word, so, in the case of proteins and nucleic acids, amino acids must be in their own specific order. For example, the hereditary disease sickle-cell anemia is caused by a single incorrect amino acid in an otherwise normal sequence of 146 amino acids of a component of the oxygen-carrying protein molecule hemoglobin, found in red blood cells (Vernon M. Ingram, "Gene Mutations in Human Haemoglobin: The Chemical Difference Between Normal and Sickle-Cell Haemoglobin," *Nature*, Vol. 180, Iss. 4581 [17 Aug. 1957], pp. 326–328).

There has been a measure of success achieved by laboratories engaged in studying the means by which amino acids, nucleotides, and other biologically significant substances could have come into existence under postulated primitive earth conditions. Scientists were able to produce the majority of small molecules that serve as building blocks for the important large molecules of the cell. For example, fifteen out of the twenty amino acids, components of nucleotides, certain vitamins, and carbohydrates have been prepared under simulated primordial conditions in the absence of oxygen. But when scientists irradiated combinations of gases that also contained oxygen, they did not obtain any biologically significant molecules (Lemmon).

Oxygen is one of our most reactive elements. It has a great tendency to combine with many other substances, among these the molecules that make up living cells (Preston Cloud and Aharon Gibor, "The Oxygen Cycle," *Scientific American*, Vol. 223, Iss. 3 [Sept. 1970], pp. 111–123). Quoting from an article on the subject: "Molecular oxygen exerts a deleterious effect on many aspects of cell metabolism, a fact difficult to account for if the first living cells had appeared in an oxygenated environment" (Lemmon).

All the experiments in which simulated primordial conditions were created and various small molecules manufactured were performed in the absence of oxygen for reasons just mentioned. Newer studies of our atmosphere, however, are forcing scientists to reconsider the validity of their model of the "primordial Earth."

It has been known for some time that, in the upper portion of Earth's atmosphere, molecules of water are shattered by strong ultraviolet radiation from the sun. The eventual products of this reaction are hydrogen and oxygen. Hydrogen, being lighter than air, escapes the atmosphere of Earth, whereas oxygen remains. Initial estimates of the amount of oxygen produced in this fashion were so low as to rate the entire phenomenon insignificant (Lloyd Viel Berkner and Leon C. Marshall, "Limitation on oxygen concentration in a primitive planetary atmosphere," *Journal of the Atmospheric Science*, Vol. 23, Iss. 2 [1 March 1966], pp. 133–143). Improved calculations, however, by R. T. Brinkman, of the California Institute of Technology, indicated that this process could produce thirty-two times the amount of oxygen found in our atmosphere over the postulated evolutionary period. Moreover, this author found that a minimum of one fourth of the present atmospheric level of oxygen should have been present for more than ninety-nine percent of Earth's supposed 4.5 billion years of evolutionary history (Robert T. Brinkmann, "Dissociation of Water Vapor and Evolution of Oxygen in the Terrestrial Atmosphere," *Journal of Geophysical Research*, Vol. 74, Iss. 23 [20 Oct. 1969], pp. 5355–5368).

The results of experiments performed during the *Apollo 16* mission appear to substantiate the calculations of Brinkman. The photographs from this mission revealed that the earth is indeed surrounded by a very substantial cloud of hydrogen extending more than 40,000 miles into space (George R. Carruthers and Thornton Page, "Apollo 16 Far-Ultraviolet Camera/Spectrograph: Earth Observations," *Science*, Vol. 177, Iss. 4051 [1 Sep. 1972], pp. 788–791). The source of this hydrogen is believed to be the water vapor in our atmosphere. Correspondingly large amounts of oxygen must have been funneled into the earth's atmosphere.

For obvious reasons, most evolutionary theorists believe Brinkman overestimates the amount of oxygen produced by the splitting of water in our upper atmosphere. If they accepted Brinkman's calculations, it would virtually nullify all existing theories of biochemical evolution.

Leigh Van Valen, member of the committee on evolutionary biology at the University of Chicago, also challenged the idea of a long, slow buildup of oxygen in our atmosphere. He showed that oxygen at about the same level as in today's atmosphere has been present for a much longer period than is generally considered by most evolutionists and that the "cause of the original rise in oxygen concentration presents a serious and unresolved quantitative problem" (Leigh Van Valen, "The History and Stability of Atmospheric Oxygen," *Science*, vol. 171, Iss. 3970 [5 Feb. 1971], pp. 439–443)—at least for biochemical evolutionists. Regardless of the position the evolutionary-minded scientist takes, it appears to be no longer possible for them to evoke long periods of oxygen-free atmosphere in postulating the early events of biochemical evolution. Thus, the theory of biochemical evolution remains very much in the "theory" state.

If the first living cell originated out of a colossal accident and various forms of life evolved from it during a constant struggle for survival, then there may be no greater purpose to life than simply to exist. A disturbing but logical application of this thought is to devote one's energy toward getting the most out of the few years allotted to humans. For individuals and nations it would be understandable—even "moral"— to exploit the weaker if the cardinal rule of existence was the survival of the fittest.

By contrast, the biblical description of origins implies a call to love our fellow humans as ourselves. (See, for example, Matt. 22:39.)

> *If the first living cell originated out of a colossal accident and various forms of life evolved from it during a constant struggle for survival, then there may be no greater purpose to life than simply to exist.*

Bible-believing Christians have often been accused of ignoring the facts of science. It appears, however, that as knowledge about our universe continues to increase, the biblical version of our origin becomes more and more credible and intellectually respectable as a reasonable alternative to evolution.

CHAPTER 3
The Billion Dollar Question

His was the billion-dollar question: "Is there life on Planet Mars?" It was not a particularly new query. Probably it had been asked for hundreds of years. But previous generations could only guess at the answer. Now we can send instruments to Mars to make direct measurements and discover the real answer.

Mars is the seventh in size among the planets of the solar system. Its diameter is more than fifty percent of Earth's diameter, but its mass is only ten percent of our planet's mass. Once every twenty-six months we come as close to Mars as 35 million miles, and, at such times, the planet glows with a reddish hue and is brighter than the brightest star. Viewed through a telescope, the Martian surface appears reddish-orange, with irregular greenish patches and two glistening white polar caps. A number of astronomers, beginning with the Italian Schiaparelli in 1877, reported thin, artificial-looking lines, or "canals," traversing the planet. To many, the greenish regions suggested the existence of vegetation, and the "canals" hinted at the intriguing possibility of intelligent life, or at least life, on Mars.

Of all Earth's neighbors in the solar system, Mars is considered to be the most hospitable to life as we know it. The temperature of its surface is never excessively hot, never more than 86°F (30°C); and its average surface temperature is only 122°F (50°C) colder than on Earth. Martian conditions are less severe than those of the boiling hot springs of Yellowstone National Park or of the water 30,000 feet deep in the Pacific Ocean, and microorganisms have been found thriving in both of these locations. So, why should there not be life on Mars as well?

During the 1960s and 1970s, a number of spacecraft launched by the United States traveled to within a few thousand miles of Mars and surveyed the planet for possible landing sites. The pictures revealed a desolate, comparatively featureless planet with craters, sand dunes, and ridges reminiscent of the lunar surface.

Telemetric data also indicated the presence of an atmosphere much thinner than ours, consisting mostly of carbon dioxide, some water vapor, carbon monoxide, oxygen, and atomic hydrogen. No traces of nitrogen, ammonia, or methane were found by the Mariner space probes.

Close-up photographs did not verify the existence of canals on Mars, nor were explanations obtained for the supposed green areas of earlier, earth-based observations. The absence of nitrogen and ammonia and the low water vapor content of the Martian atmosphere discouraged speculation about the possibility of life there.

The current popular theory of chemical evolution assumes that the initial processes that eventually led to the appearance of primitive life forms began in the atmosphere of the planet. The necessary ingredients of such a "life-producing" atmosphere were water vapor and carbon and nitrogen-containing gases. Under the influence of ultraviolet radiation (or perhaps other energy sources), the components of this atmosphere then combined to form biologically significant compounds. Amino acids, simple sugars, and fats produced in this manner in the atmosphere collect on the surface of the planet. Given sufficiently long periods of time, these simple substances—the theory proposes—assemble themselves into proteins, complex sugars, nucleic acids, membranes, and eventually into living entities.

Laboratory experiments have been performed in which various mixtures of gases have been irradiated by ultraviolet or other types of radiation, and simple biologically important substances have indeed formed in this manner. These results have encouraged evolutionary theorists to elevate their theories to the level of dogma. In essence, they have been saying that given the proper ingredients of a planetary atmosphere and the proper surface temperature and surface composition, plus a few billion years, it is inevitable that life will appear on such a planet.

Mars would have been a perfect test case for the correctness of these theories were it not for the reported absence of nitrogen-containing substances in the Martian atmosphere. Nevertheless, early in 1971, scientists at the Jet Propulsion Laboratory in Pasadena, California, exposed a gaseous mixture of carbon dioxide, water vapors, and carbon monoxide to ultraviolet radiation, and they observed the formation of formaldehyde, acetaldehyde, and glycolic acid. These organic molecules could potentially convert into biologically important substances if they interacted with the nitrogen of the Martian soil. Thus came the announcement from Pasadena that the existence of primitive life on Mars was possible.

Shortly before launching the missions, Stanley Miller, a leading scientist in the field of chemical evolution, wrote: "We are confident that the basic process [of chemical evolution] is correct, so confident that it seems inevitable that a similar process has taken place on many other planets in the solar system.... We are sufficiently confident of our ideas about the origin of life that in 1976 a spacecraft will be sent to Mars to land on the surface with the primary purpose of the experiments being a search for living organisms" (Stanley L. Miller, quoted in *The Heritage of Copernicus*, 1974, Jerzy Neyman, editor, p. 328).

This development paved the way for an all-out effort to find life on Mars. Several years of planning and instrument building and the expenditure of $1 billion followed. Then, in 1975, two spaceships with no crews were launched from the Kennedy Space Center toward Mars. Each of the 7,700-pound Viking units contained a Mars-orbiting satellite and a lander vehicle. The orbiter portion was equipped with two-way communication facilities, computers, solar-energy panels, jet-propulsion engines, and reservoirs of propellant fuel. The lander, a hexagonal-shaped, three-legged aluminum structure, housed computers, power units, cameras, and scientific instruments.

Cruising through space at approximately 30,000 miles per hour, the first spaceship touched on the Martian surface 335 days after launching.

Prior to landing, the spacecraft was placed in orbit around Mars. Potential landing sites were photographed by the orbiting vehicles for a closer look, and it was then that space scientists realized that the terrain of the initially selected site was too hazardous for a soft landing. Four weeks of intensive photographic search followed before a suitable spot was located on the Chryse Planitia basin. Then on July 20, 1976, at about 4:00 p.m. local Mars time, the *Viking I* lander successfully touched down close to the designated site and began transmitting data back to earth. A month and a half later, *Viking II* lander was also placed on Mars, at a region known as Utopia Planitia, some 4,600 miles from the location of the first robot.

The lander vehicles had been designed to conduct significant chemical and biological experiments to test for the presence of life. Based on our experience with living matter on Earth, it is safe to generalize that living matter is relatively rich in the elements carbon and hydrogen, while, in nonliving matter, oxygen is relatively abundant. Among the instruments aboard the *Viking I* and *Viking II* landers were combinations of gas chromatograph-mass spectrometers. These units could analyze the molecular and atomic components of gaseous substances.

A mechanical arm scooped up a small amount of Martian soil and placed it in an inner chamber. The soil was heated to 392°F (200°C) to drive off any relatively volatile substances, and the vapors were analyzed. Only water vapors were detected, believed to come from hydrated minerals in the soil. Next, the soil was heated to 662°F (350°C) and then to 932°F (500°C). At these temperatures all carbon-containing molecules break down to gaseous fragments, suitable for analysis by the gas chromatograph-mass spectrometer units. The results of these experiments by both Viking units were negative. Within the sensitivity of these instruments, which was ten parts per billion, no carbon-containing substances were found in the Martian soil. By contrast, surface samples from the biologically destitute regions of Antarctica have yielded some organic matter when similarly treated, at levels of several thousand parts per billion.

The subsequent experiments, designed to probe the biological activities of the Martian soil, were anticlimactic, though the results were very surprising to scientists. One of these experiments tested the ability of the Martian soil to convert radioactively labeled carbon dioxide and carbon monoxide to larger carbon-containing substances both in the dark and in the presence of light. This is routinely done by some earthbound microorganisms and by all plants. Another experiment examined the ability of Martian soil to break down and metabolize compounds labeled with radioactive carbon. A third type of experiment consisted of monitoring the release of oxygen and other gases from soil samples, as they were incubated in a complex growth medium.

The results obtained were extremely puzzling in view of the total absence of carbon-containing substances, thought to be indispensable components of living organisms. All the experiments yielded positive data, which, in Earth-based laboratories, would have been interpreted as unequivocal proof of biological activity and of the presence of life.

First, the Martian soil converted carbon dioxide to larger organic compounds to a slight extent. This ability of the soil was destroyed when the sample was heated prior to the addition of carbon dioxide. The Martian soil could also break down complex organic molecules to carbon dioxide, and pretreatment of the soil with heat destroyed this capacity of the soil as well. Third, when soil samples were moistened with water vapors, a rapid release of significant quantities of oxygen was noted. Along with this, oxygen, carbon dioxide, carbon monoxide, nitrogen, and argon also evolved. Preheating the soil before the addition of water abolished the observed phenomena.

After a review of the results, the preliminary scientific opinion was that in view of the absence of carbon-containing substances, all these data can be best explained by purely chemical reasoning. It was postulated that extensive ultraviolet radiation of the sun interacted with the inorganic minerals of the Martian surface to create exotic and highly reactive substances that were responsible for the observed results of the biological experiments. But attempts to duplicate the Viking data in Earth-based laboratories were unsuccessful. The first interim report by project scientists concluded rather optimistically, "Thus, despite all hypotheses to the contrary, the distinct possibility remains that biological activity has been observed on Mars."

In July 1977, Cyril Ponnamperuma's laboratory at the University of Maryland reported the results of experiments in which all the positive results of the Viking biological experiments had been duplicated using metal peroxides or the iron oxide, hematite, exposed to ultraviolet radiation in the presence of carbon dioxide (Cyril Ponnamperuma, Akira

Shimoyama, Masaaki Yamad, Toshiyuki Hobo, and Ramsay Pal, "Possible Surface Reactions on Mars: Implications for Viking Biology Results," *Science*, Vol. 197, Iss. 4302 [29 July 1977], pp. 455–457). These findings provided the basis for the most reasonable explanation of all the observations.

Late in 1977, project scientists of the National Aeronautics and Space Administration and of the Space Board of the National Academy of Sciences met to confer on the results of the Viking probes, with particular emphasis on the chemistry and biology of the Martian surface. After a thorough review of the data, the consensus was that Mars lacks every form of life, including microorganisms, and the search for life on that planet may be abandoned. Gerald Soften of NASA's Langley Research Center was quoted: "I may have been prepared for the lack of life on Mars, but it never occurred to me that there would be no organic chemistry as well. Before the landings, most of the scientists at this meeting would have expected to find some sort of microorganisms in the Martian soil, but now I think just about everybody would have to say that, given the data we've received, it's highly unlikely that there is any life at all on Mars."

> "I may have been prepared for the lack of life on Mars, but it never occurred to me that there would be no organic chemistry as well."
> — Gerald Soften of NASA's Langley Research Center

Through decades of continual reiteration, prominent scientists have attempted to persuade the public to accept evolutionary theories as historical facts. Science and its practitioners have earned the confidence of the public by many novel discoveries and startling technological breakthroughs.

The theories of chemical evolution are said to be valid not only for Earth but for any planet in the universe that possesses the necessary raw materials and a continuous supply of energy from a nearby star. Mars admirably fits this category. Simulated Martian environment in the laboratory produced organic molecules with potential biological significance. Successful laboratory simulations of primordial synthesis of biologically important substances serve as the foundation for chemical evolutionary theories. The Viking results clearly show that the laboratory synthesis of these substances in a simulated environment does not necessarily mean their actual accumulation on a planetary surface. In the case of Mars, highly reactive peroxides in its soil quickly degrade any organic molecule that may form in the Martian atmosphere. Prior to the Viking experiments, no one had seriously worried about the effect of unceasing ultraviolet radiation on exposed inorganic mineral surfaces. Now the evidence points to the creation of a chemically highly reactive type of matter that can confound the best schemes of chemical evolution.

Was it worth $1 billion to learn that there is neither life nor organic chemistry on Mars? From a creationist's perspective it was priceless to see that chemical evolution did not operate on our planetary neighbor. If chemical evolution is invalid for Mars, by inference it is also invalid for Earth.

CHAPTER 4

A Creationist's View of the Solar System

Earth is part of the solar system, a complex array of cosmic bodies. Traveling in nearly circular orbits around the sun are eight planets, five dwarf planets and more than 150 satellites, or moons. Unnumbered asteroids and comets are also part of our cosmic backyard. Beyond the solar system is an enormous vacuous space, stretching in excess of 25 trillion miles (4.37 light-years) in every direction. At this distance is our nearest neighbor in the Milky Way galaxy, the Alpha Centauri Planetary system.

Beginning on October 4, 1957, with the launch of Sputnik 1, the first earth orbiter satellite, the past 66 years have witnessed the launching into space of more than two hundred space probes to investigate the Moon, the Sun, Mars, Venus, Mercury, Jupiter, Saturn, and Comet Halley, and even landing a probe on comet 67P in November of 2014, mankind setting foot on the Moon, and robotic laboratories looking for evidence of life on Mars.

In 1990, the Hubble space telescope was placed in low earth orbit. With its 7.9-foot mirror, in the almost total absence of background light, it has been sending us unprecedented images of deep space in the visible light spectrum. This instrument also captures images in the near ultraviolet and infrared regions. Terrestrial telescopes cannot obtain such data as our atmosphere strongly blocks these wavelengths.

Of the eight planets in the solar system, the inner four are smaller and are composed of rocks and metal. The four outer planets are the "gas giants" Jupiter and Saturn, which contain mostly hydrogen and helium. Uranus and Neptune are made from frozen water, ammonia, and methane.

Besides our moon and two smallish satellites around Mars, there are no additional bodies in the inner solar system. By contrast, there are more than 150 satellites, or moons, of varied sizes that orbit the gas giants.

Between the orbits of Mars and Jupiter is an "asteroid belt" comprised of a very large number of irregularly shaped asteroids. The sizes of these asteroids range from the dwarf planet Ceres, with a diameter of 950 kilometers, down to particles of dust.

> The best chance of finding extraterrestrial life forms was on Mars.

Images from the surfaces of the Moon and Mars show only barren, desolate, rock-strewn landscapes. Photographs of Earth, the Moon, Mercury, and Mars from orbiting spacecrafts display numerous impact craters. It also appears that there are no living organisms on any of the planets and moons outside of Earth. The best chance of finding extraterrestrial life forms was on Mars. But in 1976 the two Viking robotic laboratories that landed on the red planet determined that not only were there no life forms there, but the landing crafts' mass spectrometers, at their sensitivities of ten parts per billion, could not find any organic substances in the Martian soil.

According to standard textbook explanation, our solar system came about by the gravitational collapse of a gigantic cloud of dust and gas some 4.6 billion years ago. This hypothesis, a modern version of the nebular theory, initially proposed in the 18th century by Emanuel Swedenburg (1734), Immanuel Kant (1755), and Pierre-Simon Laplace (1796), has grave difficulties.

The hot gaseous nebula, the source of our solar system, supposedly originated from the "Big Bang." But hot gases expand rather than collapse.

A Creationist's View of the Solar System

The mass required to oppose such an expansion was calculated to be 100,000 times the mass of the Sun—an impossibility!

Nevertheless, if somehow the spinning solar nebula was cold at the time of its collapse, then its angular momentum should have been transferred to the Sun, which has 99.8% of the mass of the solar system. In reality, it is the planets that have 98% of angular momentum of the solar system.

In a solar system originating from a homogeneous spinning cloud of gas and dust, it would be expected that every planet would rotate in the same direction. But planet Venus rotates in the opposite (or retrograde) direction of the other planets. Further, the composition and general properties of the planets would be expected to be at least similar. Instead, each inner and outer planet is unique.

Despite its demonstrable failures, this 200-year-old nebular theory of the solar system's origin is still preferred by the secular world to the 3,500-year-old transcript of the Creator's spoken words: "For in six days the LORD made heaven and earth ..." (Exod. 20:11).

Nonetheless, there are untold millions who continue to believe the truthfulness of the biblical record. But what are they to make of the current picture of the solar system? Will they just ignore the avalanche of new data from space explorations of the recent decades? Or shall their belief in the truths of the Bible give rise to alternative explanations of cosmic realities?

The Bible reveals a vital principle regarding the Creator's activities. Undergirding every act of God is a far-reaching, loving purpose, stemming from infinite wisdom.

Regarding the creation, the prophet Isaiah wrote: "God himself that formed the earth and made it; he hath established it, he created it not in vain, he formed it to be inhabited" (Isa. 45:18). Indeed, earth is teaming with millions of different types of organisms, so much so that there may not be a square inch of sterile surface anywhere on the globe!

But what about the other planets of the solar system? Their lifeless, barren states remind us of the state of Earth after the first day of the creation: "And the earth was without form, and void ..." (Gen. 1:2). Were they created "in vain" (Isa. 45:18)?

The creation week, described in the first two chapters of Genesis, marked the emergence of not only Earth, Sun, Moon, but of the entire solar system. This was going to be the residence of a new kind of created beings, whose distinction was that they were made in the image of God (Gen. 1:26). When Jesus explained that "the Sabbath was made for man" (Mark 2:27), He implied that the solar system was created for us.

The first two chapters of Genesis describe the step-by-step renovation of the barren, lifeless earth into a beautiful, Edenic habitation for Adam and Eve and their descendants. "As the earth came forth from the hand of its Maker, it was exceedingly beautiful. Its surface was diversified with mountains, hills, and plains, interspersed with noble rivers and lovely lakes; but the hills and mountains were not abrupt and rugged, abounding in terrific steeps and frightful chasms, as they now do; the sharp, rugged edges of earth's rocky framework were buried beneath the fruitful soil, which everywhere produced a luxuriant growth of verdure" (Ellen G. White, *Patriarchs and Prophets*, p. 44).

This transformation was done in the sight of a thrilled audience of created beings, who shouted with joyous excitement in seeing God in creative action (Job 38:7). A keen awareness of the Creatorship of the LORD is the foundation of our sense of well-being and happiness. This is why sinless Adam and Eve were admonished to observe weekly the memorial of creation (Gen. 2:3).

The Garden of Eden was intended to be a model to be copied by the children of unfallen Adam (Ellen G. White, *Education*, p. 22). Instead of cities, the LORD wished humanity to live in gardens. Eventually humanity would have occupied all available space on earth and then the LORD would have converted the inner planets of the solar system—as well as some of the moons of the gas giants—to places fully habitable by humans. This work would have occurred in full view of humanity, so that we, too, could have witnessed the LORD's creative prowess.

The hydrogen and helium contents of planets Jupiter and Saturn suggest that they may be small, unignited suns. Ignited, they could furnish light and heat to perhaps six of the largest moons circling the outer planets of the solar system.

The barren appearance of the inner planets, the lack of ignition of planets Jupiter and Saturn may be understood by creationists as the result of mankind's rebellion and the interruption of God's original plan.

The presence of asteroids, comets, dwarf planets, small satellites (or moons) and impact craters should also trouble creationists. Are these leftover material from the original creation of the solar system for which the LORD did not have any use? This would be completely out of character for God. The LORD does not miscalculate, and He does not waste resources.

The Creator designed Earth's ecosystem so that every single naturally-made organic substance is recycled. Soil and marine microorganisms can even remedy spills from oil wells. Grand cycles of carbon, nitrogen, and sulfur in nature ensure the maximum usefulness of every atom.

The Bible refers to "war in heaven," an open conflict between the forces of God and those of Lucifer. The asteroids (and their close relatives, the comets), the dwarf planets, the small moons, and the impact craters do not bear the Creator's signature. Their existence fits Jesus' description: "An enemy hath done this" (Matt. 13:28).

But when did all this happen? The Bible does not refer to any cosmic events that could have generated the undesirable components of the solar system. Lucifer and his angels were expelled from heaven before the creation of earth and the solar system. We know this because Satan, the enemy, appeared in the Garden of Eden.

It is likely that the undesirable modification of the solar system occurred after the Flood. During the cataclysmic, worldwide Flood, Satan was forced to remain on Earth, and he feared for his survival (Ellen G. White, *Patriarchs and Prophets*, p. 99). It may be that in retaliation for his ordeal, the forces of darkness destroyed an entire planet in the solar system. The current asteroid belt would be the remnant of this planet, its orbit approximating that of the former planet. The bulk of planetary fragments would have been sucked into the Sun by gravitation, but not before pockmarking with craters every inner planet and the Moon. Fragments flying away from the Sun may also account for the dwarf planets, small moons, and even comets.

A few years ago two scientists presented a computer model that suggested that at one time there was a "Planet V" between the orbits of Mars and Jupiter (John E. Chambers and Jack L. Lissauer, 33rd Lunar and Planetary Science Conference, Houston TX, March 11–15, 2002). They postulate that the orbit of Planet V was unstable and the planet was destroyed when it strayed into the Sun.

It is very likely that this broken solar system will be swept away when the LORD creates a new heaven and a new earth (Rev. 21:1). What will the new solar system be like? All indications are that we will have an answer to that question soon.

CHAPTER 5

The Slum of the Universe

Our earth and its immediate cosmic neighborhood is the slum of the universe. One could crisscross the vast expanse of the cosmos, traveling billions of light-years, without encountering another planetary system with a star that is orbited by eight planets; three rocky wastelands, four inert gas giants, and one solitary blue globe covered with living organisms. The inhabitants in charge of this planet are ravaged by sickness and embroiled in violence and wars, and their life spans are generally less than a hundred years. Our planet is also the most primitive and technologically backward society in the universe. But superseding all these miseries is the overwhelming ignorance of the causes and remedies of our condition.

At creation, earth was a noteworthy member of the vast array of inhabited planets. "As the earth came forth from the hand of its Maker, it was exceedingly beautiful. Its surface was diversified with mountains, hills, and plains, interspersed with noble rivers and lovely lakes; but the hills and mountains were not abrupt and rugged, abounding in terrific steeps and frightful chasms, as they now do; the sharp, rugged edges of earth's rocky framework were buried beneath the fruitful soil, which everywhere produced a luxuriant growth of verdure" (*Patriarchs and Prophets*, p. 44).

Our planet and its plants, flowers, animals, birds, and fish were brand new, one-of-a-kind constructs at the time of the creation. We know this because the creation account in Genesis 1 mentions that the Creator several times double-checked His handiwork to make sure that everything turned out as originally planned. If our world were a carbon copy of previously produced planets, there would have been no need for such quality control.

Adam and Eve, the jewels of the creation and the masters of the world, were also brand new additions to the order of created beings. "All heaven took a deep and joyful interest in the creation of the world and of man. Human beings were a new and distinct order" (RH, Feb. 11, 1902, Art. A). Created in the image of God, humans were equipped to be new links between the Creator and the rest of the creation.

The all-surpassing beauty of the newly created earth, the high privilege of being made in the image of the Creator, and the wonderful garden-home, prepared expressly for Adam and Eve, were some of the magnificent gifts the LORD showered on them. These were the tangible evidences of the deep love of God for our first parents.

We envy Adam and Eve for their strolls through the Garden with their Maker in the cool of the day. What it must have been like to see the LORD's smiling face, to hear His melodious voice addressed just to them! We imagine how their hearts raced as they spent quality time with the Being who fashioned their every cell, who knew all their thoughts and feelings, who spoke to them life-giving words. It was on such an occasion that they learned the story of their origin from the Creator Himself and from His angels (*Patriarchs and Prophets*, p. 83).

Had the first pair fully understood that these honors were contingent on their continuous, perfect obedience, perhaps their history and that of mankind would have turned out differently. As it happened, their disastrous transgression banished them from the visible presence of the Creator, and our world was cut off from direct communion with the LORD as well as from the vast communities of the cosmos. Humans could no longer benefit from the accumulated knowledge and wisdom of the universe. We had to reinvent the wheel and everything else.

In the absence of direct and immediate heavenly influences, mankind drifted onto uncharted and dangerous moral landscapes. "Not desiring to retain God in their knowledge, they soon came to deny His existence. They adored nature in place of the God of nature. They glorified human genius, worshiped the works of their own hands, and taught their children to bow down to graven images" (*Patriarchs and Prophets*, p. 91).

The LORD bore patiently with mankind's downhill moral slide for more than fifteen hundred years. But eventually the low point was reached. "And GOD saw that the wickedness of man was great in the earth, and that every imagination of the thoughts of his heart was only evil continually" (Gen. 6:5).

The Great Flood destroyed all life on land. Its aftermath altered the planet's surface for the worse, paralleling humanity's moral brokenness. But why wasn't earth simply obliterated at this time? The experiment of humanity's rehabilitation clearly had failed. What was to be gained by prolonging mankind's rebellion against heaven? The onlooking universe was puzzled.

Then an unexpected event of such magnitude occurred that it took away the collective breath of unfallen beings. The majestic Creator of the universe inserted himself into our history by becoming a human infant! The life and death of Jesus Christ is the most monumental event in the entire history of the universe, worthy of study throughout eternity!

Here was the reason God did not annihilate earth in its totality following the initial sin of Adam and Eve. "For God so loved the world, that he gave his only begotten Son, that whosoever believeth in him should not perish, but have everlasting life" (John 3:16).

Whatever plans the LORD had for the solar system before Adam and Eve's transgression were clearly abandoned. It has been left in a chaotic state to this day, a reflection of mankind's rebellion (George T. Javor, "A Creationist's View of the Solar System," *Dialogue*, Vol. 27, No. 2 [2015], pp. 13–15).

The onlooking universe eagerly watched what would happen next. At the cross, Lucifer's rebellion came to a climax. There he was unmasked as the willing killer of his own Creator! If there was any sympathy reserved for Lucifer among the unfallen beings before, it vanished.

Now it was just a matter of heralding throughout the entire earth the best possible news: God Himself suffered the punishment for mankind's sin, and a way was opened for humanity's existence through eternity! If anyone accepts God's sacrifice on his or her behalf, he or she is saved! The Savior's life and death on the cross established the escape route from the slum.

Who would not accept this fantastic gift? The onlooking universe anticipated a rapid conclusion of this chapter of earth's history and the complete rehabilitation of mankind into the community of unfallen beings.

Alas, most of humanity worshipped idols, and they were in no position to understand—much less appreciate—God's incredible sacrifice. They had no clue that, in the absence of this great sacrifice, earth could not exist much longer. Subsequent centuries witnessed the growth of Christianity, but the gospel message never covered the globe.

Two thousand earth-years later the onlooking universe is still puzzled at mankind's dullness of comprehension. Twenty centuries after the initial rescue of humanity from oblivion, the heralds of the Good News are still a woefully small minority. To make matters worse, modern science, driven by the philosophy of materialism, now has preeminence over religion.

Today's scientists are the priests at the altar of knowledge. Thought leaders look to them for reliable information about our world and the universe. The Bible, the only collection of writings containing the record of God's dealings with humanity and the plan of salvation, is set aside as unworthy of serious consideration.

Universities now teach students that the universe originated from a gigantic explosion, and everything we see around us, including the galaxies, the solar system, the earth and its biosphere, are all the result of a continuous series of fortuitous accidental events over the past billions of years. The Creator, if He exists at all, is merely an observer of the continuous evolution of the universe.

The late Steven Hawking, a preeminent British physicist, published a book in which he asserted that God did not create the universe and the "Big Bang" was an inevitable consequence of the laws of physics. He further asserted, "Because there is a law such as gravity, the Universe can and will create itself from nothing" (Stephen Hawking with Leonard Mlodinow, *The Grand Design*, 2010, p. 199).

Compared to our "wise men," the ancient pagan priests and philosophers were geniuses, as they at least acknowledged the existence of supernatural beings. The purposeful denial of the Creator and the militant promotion of evolution in the classroom condemn the world's elite to an intellectual world of perpetual darkness (George T. Javor, "Letters to Editor," *Microbe*, Vol. 3, Nov. 5, 2008; because this volume is missing from *Microbe*'s archive, the contents of this letter are reproduced at the end of article).

Our world and everything on it is here by God's design, but it is not a machine that was wound up once and is now running on its own. Nothing in the entire universe, animate or inanimate, exists independently of the great Creator. The LORD is not only our Creator but also our Sustainer moment by moment (Heb. 1:2, 3; Acts 17:24, 25; Job 34:14, 15). "Those who have a true knowledge of God will not become so infatuated with the laws of matter or the operations of nature as to overlook, or refuse to acknowledge, the continual work of God in nature. Nature is not God, nor was it ever God.… The natural world has, in itself, no power but that which God supplies.… God is the superintendent, as well as the Creator, of all things. The Divine Being is engaged in upholding the things that He has created.… It is through the immediate agency of God that every tiny seed breaks through the earth, and springs into life" (*Selected Messages*, bk. 1, pp. 293, 294).

Unfallen beings throughout the universe rejoice in being sustained by the Creator. They understand that this is the ultimate expression of His great love for His creatures. They could not imagine existence without their beloved God! By contrast, on this sin-darkened planet, the majority of humans are unaware of the glorious destiny that can be theirs. What a task for the followers of the Creator to awaken their slumbering fellow humans to this reality!

This slum of the universe, currently a battlefield between good and evil, is to be restored to a glory much greater than originally planned. It will become the most sacred part of the universe, housing the very throne of God. It will become the most sought-after place. The New Earth will be the destination of visitors from the farthest reaches of the cosmos. It may be that to accommodate them, the entire solar system will be re-created to house and feed our fellow space-tourist friends throughout eternity. It will be the privilege of the redeemed to welcome, in the New Earth, visitors from other worlds and join them in worshipping the great Creator who made everything possible.

Evolution in the Classroom

Here is the letter to the editor referred to above:

Risking the ire of the National Academy of Sciences, attention needs to be called to the irony in their current crusade against creationism in science classrooms. Sir Francis Bacon, who is credited with formulating and establishing the scientific method, was a creationist. So Were Sir Isaac Newton, Robert Boyle, Louis Pasteur, Carl Linnaeus, Michael Faraday, Blaise Pascal, Lord Kelvin, James Clerk Maxwell, Jean Louis Agassiz, Rudolph Carl Virchow, Johannes Kepler, and numerous other intellectual giants on whose shoulders stands the modern scientific enterprise. Clearly, creationism did not hinder the scientific work of these greats, rather it encouraged them to seek keener insights into the secrets of the physical realm. Permitting students to peek outside the box of evolution is hardly a dilution of science. Rather, it is granting them freedom of imagination and thought similar to what students of previous generations were allowed to have.

George T. Javor
Loma Linda University School of Medicine
Loma Linda, California

CHAPTER 6

Biblical Approaches to Biology

In contrast to those who restrict the authority of the Bible exclusively to moral and religious topics, many Christians accept the Bible as the revealed Word of God on all matters. As such, they attempt to harmonize their understanding of science with relevant biblical information.

This exercise rests squarely on the conviction that the Bible contains supernatural revelation. Without this, how could one assign such a dominant role to material written 3,500 years ago by those who were utterly innocent of any knowledge of modern science?

The integration of Bible and science is an uphill work that requires careful reading of both the Bible and scientific data. It is best done in collaboration between theologians and believing scientists.

No science requires this corrective procedure more than biology. This is not because biology is becoming the dominant science of our age, although this development gives added urgency to such work. Rather, it is because no other natural science has traveled so great a distance down an anti-biblical road. Currently, in order to accept the teachings of modern biology at face value, one has to discard, ignore, or, at the minimum, drastically reinterpret what the Bible teaches on these matters.

By way of illustration let's look at two examples of the relationship between modern biology and religion. The first is an article by Theodosius Dobzhansky published in *The American Biology Teacher* in 1973 entitled, "Nothing in Biology Makes Sense Except in the Light of Evolution" (*The American Biology Teacher*, Vol. 35, Iss. 3 [March 1973], pp. 125–129).

In his article, Dobzhansky states that he is a religious person, even a "creationist," who believes that God created and continues to create through evolution. He makes the following observations: 1. The Bible is not a primer of natural science. It treats "matters even more important, the meaning of man and his relations to God." The Bible "is written in poetic symbols that were understandable to people of the age when they were written, as well as to peoples of all other ages." 2. "Contrary to Bishop [James] Ussher's calculations, the world did not appear approximately in its present state in 4004 B.C. The estimates of the age of the universe given by modern cosmologists [are] … about 10 billion years old. The origin of life on earth is dated tentatively between 3 and 5 billion years ago; manlike beings appeared … between 2 and 4 million years ago." 3. "Anti-evolutionists fail to understand how natural selection operates. They fancy that all existing species were generated by supernatural fiat a few thousand years ago, pretty much as we find them today." 4. Despite its great diversity, there is a basic unity of life, suggesting that it arose from inanimate matter only once. If the millions of species found today were all created by separate fiat, then the Creator "deliberately arranged things exactly as if his method of creation was evolution, intentionally to mislead sincere seekers of truth." 5. Besides the biochemical universals, comparative anatomy and embryology also proclaim evolutionary origins. Examples are homologies in the skeletons and other organs of all vertebrates, the striking similarities among embryos of diverse animals, the presence of nonfunctioning gill slits in human and other terrestrial vertebrate embryos. 6. Without the light of evolution, biology is a pile of sundry facts, some of it interesting or curious but as a whole not meaningful.

The second example is from the introductory chapter of a recent textbook of college biology with the heading, "Science and Religion" (William

K. Purves, David E. Sadava, Gordon H. Orians, and H. Craig Heller, *Life, the Science of Biology*). Here is a portion of this material:

"Creation science is not science. Science begins with observations and the formulation of testable hypotheses. Creation science begins with the unsubstantiated assertion that Earth is only 4,000 years old and that all species of organisms were created in approximately their present forms. This assertion is not presented as a hypothesis from which testable predictions are derived. Advocates of creation science do not believe that tests are needed because they assume the assertion to be true.

"In this book we present evidence supporting the hypothesis that the Earth is several billion years old, that today's living organisms have evolved from single-celled ancestors.... All of this extensive scientific evidence is rejected by proponents of creation science in favor of a religious belief held by a very small minority of the world's population. Evidence gathered by scientific procedures does not diminish the value of the biblical account of creation. Religious beliefs are not based on falsifiable hypotheses, as science is; they serve different purposes, giving meaning and guidance to human lives. The legitimacy of both religion and science is undermined when religious belief is called science" (quoted by George T. Javor in *Biblical Approaches to Biology*).

These comments, which suggest that the Bible and religion in general have no useful input to science and that the applying of religion to science will destroy the effectiveness of science, approximate the official stance of most scientific organizations on this matter.

Without arguing the specifics at this point (and we will revisit some of these later), one is struck by the caricature-like, stilted characterization of the creationist's position by the deliberate blurring of the differences between facts and interpretations of facts. Evolution is presented as a single monolithic concept, and science is defined in such a way as to preclude any revelatory input.

In reality, science is about explaining how everything around us operates. Conducting science may begin with observation, but even that is done with some theoretical framework in mind. However, when students learn science, they are given information gathered by previous generations of scientists. The importance is the validity of the information, not its source. Creationists maintain that just because scientific information was obtained supernaturally by revelation rather than by experimentation, it does not diminish its value. On the contrary, having faith in the Source of the information makes that information superior to any experimentally derived scientific datum or interpretation.

Scientific data and their interpretations are not equivalent in value to biblical revelation about nature. Given that harmony must exist between the two, in case of conflict, biblical revelation must have supremacy.

> While the Bible is not a primer on science, it does contain information of great relevance to science.

While the Bible is not a primer on science, it does contain information of great relevance to science. This information is not falsifiable by testing; it can be accepted and utilized as foundational material, or it can be rejected. The same may be said of evolutionary theories, in that if one version is shown to be incorrect another variation of it is constructed. Contrary to Dobzhansky's assertions, no version of evolutionary theory is compatible with the Bible. The clash is not between science and the Bible but between evolutionary science and the Bible.

Biology, the study of life, rests on the pillars of physics and chemistry. Modern biology strongly overlaps chemistry; therefore, it seems appropriate to enter a discussion of biology between these two pillars.

Assuming that, when the LORD began to create the Earth, He did not use preexisting matter, rather, from Einstein's equation $E = mc^2$, we surmise that the Creator converted some of His energy into matter. The mass of Earth is an estimated 6×10^{21} metric tons (or 6×10^{27} grains). This quantity is calculated on the basis of the mass (m) of the Earth being 6×10^{27} grains, the speed of light (c) being 3×10^{10} cm/sec, the energy content of the Earth's mass (E) being 5.4×10^{24} grains times cm^2/sec^2, a joule being 10^7 ergs, an erg being cm times a dyne, and a dyne being grains times cm/sec^2.

The relationship between the Creator and the physical matter of the universe needs to be better understood. There can be no question about the ownership of matter, but is there more here? Is it too far-fetched to suggest that, since matter is a stable form of some of the Creator's energy, He has the capacity for absolute control over the inanimate world to the extent that He is able to track every atom?

The saying of Jesus that "the very hairs of your head are all numbered" (Matt. 10:30) perhaps can be reformulated as "every atom of your being is numbered." It is not that the LORD manipulates us through our atoms. Rather, the Creator is aware of every atom He created and has the ability to use them any way He wishes. This insight helps us appreciate how the Creator could multiply loaves and fishes, calm the Sea of Galilee, or command Lazarus to walk out of his grave.

With the creation of matter, the LORD brought into existence a universe that is at least forty orders of magnitude larger than the smallest object within it. The universe is estimated to have a diameter of 93 billion light-years or 8.8×10^{23} kilometers (in which a light-year is 9,460,730,472,580.8 kilometers). The nucleus of an atom has a diameter of 1×10^{-12} centimeter. Thus we have a 10^{40} range between the atomic nucleus and the diameter of the universe. New dimensions were created that could be populated with living beings. From the existence of radioactive elements, we know that the matter of our world is of finite age. Had matter been in existence forever, there would be no radioactive elements. Assuming that at the birth of matter there were only parent isotopes present (an assumption currently used by mainline science), it would seem that the matter of our Earth came into existence some four to five billion years ago.

A literal reading of the biblical account of creation and of the subsequent history of humanity does not readily allow for such an enormous span of elapsed time. To be sure, some Christians squeeze billions of years between verses 1 and 2 of Genesis chapter 1. But this contortion of the biblical text has a price. Now, the word "creation" can refer only to the re-organization of a preexisting, "formless" planet and the creation of living organisms. The words of the LORD etched in stone, "In six days the LORD made the heaven and earth, the sea, and all that in them is, and rested the seventh day" (Exod. 20:11), lose their potency if indeed the creation of the heavens and the earth began 4.6 billion years ago.

From the narratives found in Genesis 1 and 2, it is evident that many of the created entities—the trees in the Garden of Eden, the animals, Adam and Eve, etc.—were all brought into existence with apparent ages. It is therefore logical to assume that each of the 100 or so different elements, out of which everything was made at creation, contained their complements of isotopes, including some daughter elements of radioactive isotopes. Contrary to Dobzhansky's charge, this would not be trickery or game-playing on the part of the Creator, since the Creator personally briefed the first man as to his origins.

From the biblical context, such as Job 38:7 ("the sons of God shouted for joy" at creation), it may be reasoned that Genesis 1:1 refers to the formation of our planet and its immediate support system, including perhaps the solar system. It seems that the Earth was created as a preexisting, older universe.

Living organisms first appeared on the third day of creation week in the form of robots or machines, also known as plants, which are the connecting link between Earth and its power source, the Sun. Without plants, the energy of the sun could warm the planet but could not nourish it. It is the green solar panels of the plants that capture a portion of the electromagnetic radiation of the sun and utilize it to split water into hydrogen and oxygen. The released oxygen benefits all other organisms that respire air, and hydrogen is used by the plants to reduce carbon dioxide to carbohydrates.

Carbohydrates are compact, portable packages of energy that, when swallowed and metabolized, release solar energy to power the operation of the organism. The Genesis account is clear that the Creator designed plants, nuts, and fruits as the nutrients for all other organisms. There was no predation in the Garden of Eden.

Non-plant organisms were created next, all possessing nervous systems, all able to move and interact with their environment. The Creator's command to birds and marine organisms to multiply and fill their ecological niches indicates that they were equipped to adapt to their respective environments. That is, they could exist in relative isolation as well as in larger communities.

Creationists do not claim that the LORD created essentially all species found today. Species are reproductively isolated groups of organisms, existing undoubtedly within the Genesis "kinds." Studies of hundreds of fruit fly species of Hawaii, for example, revealed that the difference between them is in the order of genes on their chromosomes. The changes in gene order came about by stepwise mutations, and apparently all species were derived from one or two original species. This is an example of "microevolution," and creationists have no quarrel with it.

What creationists deny is that an organism from one "kind" is related to another "kind" of organism through a common ancestor. There is no clear agreement as to what level of taxonomy the Genesis "kind" would be assigned in the scale of phylum (the highest), class, order, family, genus, species (the lowest). One possible level would be above genus but below family (Siegfried Scherer, in *Mere Creation*, William A. Dembski, editor, p. 195).

Variation within each "kind" is implied in the Genesis account. One such "kind" is the human kind. By explaining that all humans are descendants of one pair, the implication of variability is clear. This variability among offspring gives true individuality to every being. We now know that all physical characteristics of individuals are determined by the nucleotide sequences of their genetic material. Variations among humans of the same race are caused not by mutations but by differences in the levels of gene activities.

The genetic material of the parents is transmitted with extraordinary fidelity to their offspring. The error rate of copying DNA is one such event per ten billion nucleotides copied! (Christopher K. Mathews, Kensal E. van Hole, and Kevin G. Ahern, *Biochemistry*, 2000). Thus, the laws of genetics prevent large-scale variations (changing one family into another).

Recent advances in genetics permitted the cloning of the sheep Dolly. She was genetically identical to another sheep, whose genetic material was used in the experiment. The first cloning procedure, however, was recorded in Genesis 2, when the LORD took the genetic material from Adam's bone cells and modified it appropriately to create Eve. This was done to have a kinship between the first human couple.

One of the biggest conundrums of modern evolution is the origin of life. Because of the complexity of living matter and because there are fundamental similarities among all the forms of life, it is assumed by evolutionists that all life forms originated from a single one-celled primitive organism. By contrast, the biblical account in Genesis 1 and 2 describe separate creations of plants, aquatic organisms, birds, terrestrial animals, and humans. Birds, animals, and Adam were all created from the ground, indicating a qualitative similarity between them. Indeed, biochemists find a lot of similarities in the gross biochemical composition of all living matter, from bacteria to man.

More detail is given for Adam's creation than for any other living creature—the LORD breathed into his nostrils the breath of life, and man became a living soul. A commonly held perception is that this "breath of life" is what distinguishes man from beast. However, we read in Ecclesiastes 3:19 that the same breath is in man and beast, nullifying that notion. Just because the Genesis account did not report animals receiving the breath of life from God does not necessarily mean that they, in fact, did not get it. After all, nothing is said about Eve receiving the breath of life either.

It is appropriate at this point to discuss what life is. Even though entire disciplines revolve around manifestations of life—biology, microbiology, biochemistry, and biophysics—it is difficult to find extended considerations of this subject.

All life forms with which we are familiar in science are associated with matter. Life in the operational sense is not a freestanding entity, something that can be isolated and studied. Rather, life is a description of the behavior of unique forms of matter. One definition of "life" is "the property or quality that distinguishes living organisms from dead organisms and inanimate matter, manifested in functions such as metabolism, growth, reproduction, and response to stimuli" (*American Heritage Dictionary*, Boston, 1991). The term "life" has multiple meanings, depending on the type of matter it is applied to.

Illustrating this point, imagine an unfortunate victim of a car accident. This person is alive one minute and dead the next. However, his organs (kidney, heart, liver, etc.) may be salvaged within a short period of time and transplanted into another body. The rescued organ will continue to live in its new environment. Cells from these organs may be put into an appropriate culture dish and can be maintained for extended periods of time. The life of the person (the organism) has a different meaning than the life of an organ, which again is different from the life of a cell.

When a cell is taken apart, one finds water (70 percent by weight), complex substances of proteins, nucleic acids, polysaccharides, and fat-like material (26 percent by weight), a mixture of simple metabolites (3 percent by weight), and inorganic salts (1 percent by weight). The shocking thing is that all these substances are inert and lifeless. What happened to life when we took apart the living cell?

We may mix the constituents of the cell together, but we continue to get a lifeless, inert mixture of chemicals. Having available the most sophisticated laboratory equipment and biochemical techniques is of no help. We just cannot restore dead cells to life.

This point is so central that it requires a more extensive discussion. Here, only the final conclusion is given: As the cell is disrupted, all chemical reactions within it reach their end points, called "equilibria." This is the chemical status of death.

We can imagine that, at creation, the LORD first built the necessary structures, which were at chemical equilibrium. Then, when He "breathed" into His creations, the non-equilibrium states of biochemical pathways were established, and life started. Biologists tell us that "life comes only from life." Thus, creation resulted in the igniting of biochemical chain reactions, which continue in living organisms to our day.

All hypothetical "primordial earth" scenarios, which purport to suggest how life may have sprung into existence, are bankrupt. Besides failing to show how information containing biologically relevant biopolymers could arise (a topic not covered here), they are also unable to show how the non-equilibrium states of biochemical pathways could come about spontaneously.

God blessed the man and woman and said to them, "Rule over the fish of the sea and the birds in the sky and over every living creature that moves on the ground" (Gen. 1:28, NIV). This mandate to humanity implies the call to study nature and to do some good biological study. To get human beings started, the LORD prompted Adam for some serious taxonomy: "Now the LORD God had formed out of the ground all the wild animals and all the birds in the sky. He brought them to the man to see what he would name them; and whatever the man called each living creature, that was its name" (Gen. 2:19, NIV). The Creator had the right to name the creations of His hands, but He deferred to Adam. Thus the first man was drawn into the creation process.

"God saw all that he had made, and it was very good... Thus the heavens and the earth were completed in all their vast array" (Gen. 1:31; 2:1, NIV). It does not take a great deal of sophistication to realize that nature has changed for the worse since creation. The second part of this quotation closes the door on the idea, suggested by Dobzhansky, that the LORD is still in the business of creation via evolution. Creationists are correct in insisting that creation was finished at the end of creation week. As we have seen before, the created kinds were endowed with the ability to vary and adapt within their genetic boundaries.

The discrepancy between the current biological world and the one described in Genesis 1 and 2 is evidence that Genesis 3 is also factual. The presence of destructive biological agents, viruses, and prions can be accounted for only as the works of an evil genius, out to sabotage the created world. Jesus identifies this malevolent person as the "enemy" (Matt. 13:28).

The extensive coal and oil deposits worldwide testify to the historical reality of the Great Flood (Genesis 6–8). Fossil remains of organisms give abundant opportunity for the study of pre-Flood biology. Mainline science assigns great ages to the fossil remains and uses them as evidence for the reality of evolution. In a recent publication, Ariel Roth presented a balanced discussion of these issues (Ariel A. Roth, *Origins: Linking Science and Scripture*, 1998, pp. 147–192).

Other than photoautotrophic microorganisms, all organisms in the biosphere are dependent on other organisms for their survival. Plants need nitrogen-fixing microorganisms to utilize the great abundance of nitrogen in the air. Plants are food for a large proportion of the animal world and indirectly even for carnivores. Thus, most living organisms run directly or indirectly on solar energy.

Oxygen in the air combines with the carbon and hydrogen of carbohydrates during metabolism and respiration in all organisms to form carbon dioxide and water. These molecules are in turn reformulated by plants to oxygen and carbohydrates through photosynthesis.

Microorganisms in the soil degrade dead organic substances, enabling the recycling of the elements carbon, nitrogen, sulfur, and phosphorus.

The dogma of "metabolic infallibility" states that every naturally occurring organic substance is biodegradable. By these means every organism is linked into a giant solar energy-utilizing network. Therefore, we have a seamless integration of the earth's rotation around the sun with life on our planet. Is it far-fetched to suggest that the Creator of the sun and the earth is also the Engineer who designed the solar-powered living organism?

These considerations help us appreciate the need for biochemical similarities among organisms. If we all use similar substances for our energy, carbon, and nitrogen needs, should not our metabolic machines also resemble each other? Thus we see molecular homologies among organisms. Creationists maintain that these reflect the signature of a common Designer rather than being an evidence of common ancestry.

In recent years, arguments for design in nature have resurfaced with renewed power. William Dembski introduced an algorithm using the laws of probability. Analysis of an event passes through three filters: high probability, intermediate probability, and small probability. Events of small probability are examined as to whether they were specified in advance. If they were, the event is judged deliberate, hence intelligently designed (Dembski, p. 93). Michael Behe showed that when biochemical systems are analyzed, one comes to a point called "irreducible complexity," which is the minimal essential for the function of the system. Removal of any part of such a complex system renders it useless. The presence of such irreducibly complex systems in living matter is an evidence for design (Michael Behe, *Darwin's Black Box: The Biochemical Challenge to Evolution*, 1996).

Another illustration of design in nature is the observation that, when predesigned components of manufactured goods are assembled, new functions emerge. Thus, when gears, pistons, sheet metal, wheels, and thousands of other components are appropriately assembled, the result is a car. It is possible to arrange the levels of our reality—from energy to the universe—in a hierarchical scale, each level acquiring new functionality.

A logical way to account for the new functionality at each level of increased complexity is to suppose that the universe has been designed. Living organisms fit remarkably well into this hierarchical order of reality. (See Figure 1 below.) It almost appears that this reality was designed for the sake of living organisms.

Figure 1

LEVELS OF REALITY	NEW FUNCTION
Energy	
↓	
Subatomic particles	Stabilization of energy
↓	
Atoms	Shape, substance, chemical properties
↓	
Molecules	Novel chemical properties
↓	
Cells	Life
↓	
Organs	Specialized tasks needed by multicellular organisms
↓	
Organisms	Complex life forms
↓	
Ecosystems	Localized interaction among life forms
↓	
Biosphere of Earth	Global interaction among life forms
↓	
Solar System	(relationships not clear)
↓	
Universe	(relationships not clear)

This essay intended to show that the great principle of creation as given in the Bible is still up to date. In fact, we can be confident that new discoveries will only strengthen the case for creation. "Since the book of nature and the book of revelation bear the impress of the same master mind, they cannot but speak in harmony" (*Education*, p. 128). Someday scientists will be forced to acknowledge that nothing in biology makes sense except in the light of creation.

CHAPTER 7

Teaching Biology in the Light of Creation

Biology, the study of life and of living organisms, is one of the most exciting subjects in the sciences. Here the student learns how cells and organisms function on the molecular and organismic levels. This knowledge is foundational for all health sciences and for nutrition.

However, biology taught in the secular setting has become "evolution in action," the epicenter of materialism which posits that matter is capable of organizing itself into the biosphere. It is now the orthodox tenant of biology that evolution is the engine that propels the synthesis of new species. Standard textbooks of biology use such logic to explain the existence and functioning of organisms. Predictably, biology teachers in Seventh-day Adventist schools are challenged to present their subject from the creationist's perspective.

Here are a few topics of biology and biochemistry that easily lend themselves to creationist interpretations. (Additional information about these topics may be found in standard textbooks of biology and biochemistry such as *Biochemistry*, third edition, by Reginald H. Garrett and Charles M. Grisham, Thomson Brooks/Cole Publishing, 2005.)

1. ***No living thing can survive by itself.*** Organisms in the ecosystem depend on other living entities for survival. Figure 2 shows the interdependence of all organisms. Humans and animals breathe oxygen produced by plants through photosynthesis. Plants, on the other hand, require nitrogen for their growth, which they receive with the help of special nitrogen-fixing

Figure 2. Interdependence of Organisms in the Biosphere

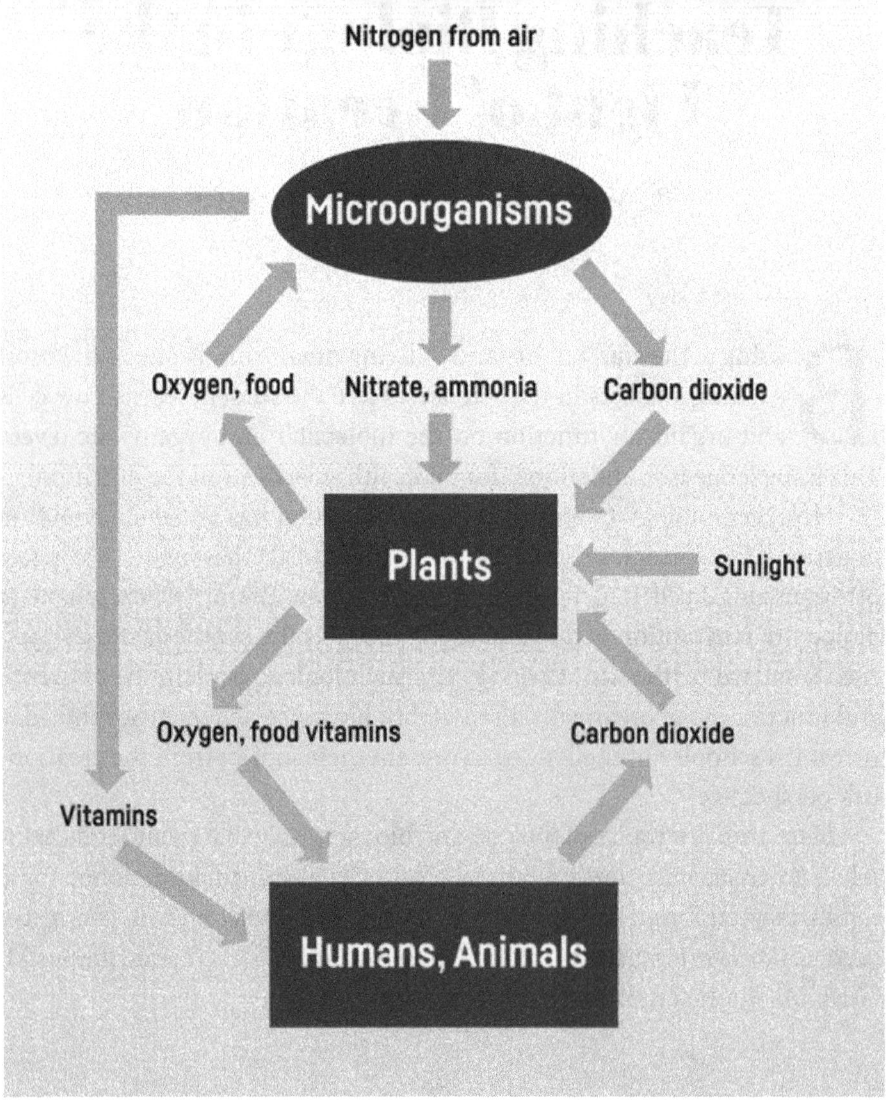

microorganisms. Intestinal microorganisms produce vitamins for our use. All this strongly suggests that the entire ecosystem had to come into existence simultaneously.

2. ***The dogma of "Biochemical Infallibility."*** There is not a single naturally occurring substance that is not metabolizable. If this would not be so, there would be large deposits of unusable organic matter everywhere, cluttering up nature and eventually causing severe shortages of the all-important carbon. As it is, the complete turnover of naturally made organic substances prevents catastrophic build-up substances in the environment. Experts predicted that the now famous BP oil spill would wreak havoc on the environment. Instead, the spilled crude oil quickly disappeared, revealing the existence of hitherto unknown marine microorganisms, which easily metabolized the spilled oil. The Creator makes provisions for every substance produced in nature. He does not tolerate waste.

3. *Enzymes catalyze every chemical reaction in living matter.* Many substances in living matter are capable of chemically interacting with each other, resulting in useless, non-biological components. Such interactions would undermine living processes. Enzymes speed up biological processes so much that there is no chance for these random chemical events to occur spontaneously. All substances in living matter are ushered along predetermined metabolic pathways as they accomplish growth, movement, metabolism, and replication. Here we see the Creator optimizing the workings of living matter.

4. *The universal existence of turnover of matter in organisms.* Biodegradative pathways are in complete synchrony with biosynthetic pathways. If either system is out of balance, the organism perishes. Even though cells expend a considerable amount of energy to build protein molecules, we now know that proteins are periodically degraded. For example, hemoglobin, the oxygen-carrying protein of the blood cells is degraded in about 120 days. Scientists eventually came to understand the important reason for the turnover of proteins. During normal metabolism protein molecules become damaged by their interaction with oxygen, rendering them ineffective and even toxic. Turnover ensures that there are no worn-out, oxidatively damaged components in living matter. Everything is "factory fresh."

5. ***The existence of apoptosis***, the programmed self-destruction of entire cells, in order to preserve the health of the tissue. Just as worn-out molecules are degraded, entire old cells from the tissue are also degraded. In the course of seven to ten years, every cell in our bodies, except for the brain cells, are replaced. Imagine the dire consequences of a failure of controls on such a system or of trying to establish it by trial and error!

6. ***The existence of topoisomerase II enzymes***, which, during replication, cut both strands of DNA while holding onto both strands. During replication, it is necessary to separate the two strands of the genetic material. However, at the point of strand separation, which is the "replicating fork," there is a tightening of the yet unseparated strands of DNA. Without cutting of the DNA strands to relieve the tension, DNA replication would come to a halt. When the topoisomerase II enzyme cuts both strands of the DNA temporarily, all that stands between the life and death of the cell is the tight grip of the enzyme of each DNA strand. Thus, in replication, the cell is pushed to the edge of death. Imagine the difficulty of such an enzyme coming into existence by trial and error.

7. ***The mechanism of the peptidyl transferase reaction on ribosomes***, which creates the peptide bonds of proteins, is identical—in the reverse direction—to the mechanism of the protein-degrading enzyme chymotrypsin. For many years, scientists did not know which of the more than fifty proteins of the bacterial ribosome catalyzed the formation of peptide bonds. Then, after the three-dimensional structures of ribosomes were determined, it was discovered that the ribosome's catalytic center is not on any of its proteins but at a special location on its ribonucleic acid. The process of peptide bond formation turned out to be identical to the exact reverse of the way peptide bonds are broken up. The surprising thing is that, in ribosomes, nucleotide bases are doing the same work as amino acid residues on the protein chymotrypsin.

This illustrates the elegance of the Creator's genius of solving biochemical problems. In this instance, the formation (and its reverse) of peptide links are promoted by selected amino acid residues in the enzyme chymotrypsin and by purine and pyrimidine residues in ribosomes.

8. ***The absolute avoidance of equilibrium*** in the hundreds to thousands of chemical reactions in living matter, in spite of the fact that every reaction

is pushed vigorously toward equilibrium by an enzyme, every chemical reaction has a beginning, a middle, and an end. At the end of a chemical reaction, equilibrium sets in and no further chemical changes happen. But life depends on continuous chemical changes in cells. If all the chemical reactions reach their equilibriums, the cell dies.

The Le Chatelier's Principle in chemistry states that once a chemical reaction reaches its equilibrium, it will not revert to a non-equilibrium state spontaneously. Cells stay alive because their chemical reactions are linked, preventing them from ever reaching equilibriums.

Chemical evolutionary scenarios all postulate that life happened spontaneously by the stepwise development of enzymes (protein or RNA molecules), which promote single chemical steps. Therefore, even if all these postulates were true, they would only result in a collection of isolated chemical reactions, all in states of equilibriums.

For living cells to form, all the chemical reactions in the cells would have to be present simultaneously, linked in states of non-equilibriums. The Le Chatelier's principle forbids the spontaneous conversion of chemical reactions from states of equilibrium to states of non-equilibrium. Therefore, all postulates that suggest a spontaneous emergence of a living cell under any conditions are an impossibility!

> *The very existence of life is an incontrovertible evidence of God's existence!*

9. We cannot reverse death or create life in the laboratory. The very existence of life is an incontrovertible evidence of God's existence!

In my thirty-seven years of teaching chemistry, biochemistry, and microbiology in the Seventh-day Adventist school system, I emphasized the LORD's creatorship, giving the LORD full credit for designing and implementing all of biology. I thought that this was the needed counterweight to the prevailing evolutionary concepts that were in currency at that time. I did not realize the unstated implication of my logic, which was as follows: the LORD created this world and the biosphere as a very complicated and very good machine, which once started, operates on its own. Aside from origins, I was teaching *a materialistic version of creationism.*

As to how things worked in biology, there was little difference between a materialist and me.

I was aware of statements from Mrs. E. G. White, such as: "It is supposed that matter is placed in certain relations and left to act from fixed laws with which God Himself cannot interfere; that nature is endowed with certain properties and placed subject to laws, and is then left to itself to obey these laws and perform the work originally commanded. This is false science; there is nothing in the word of God to sustain it. God does not annul His laws, but He is continually working through them, using them as His instruments. They are not self-working. God is perpetually at work in nature.... It is not by an original power inherent in nature that year by year the earth yields its bounties and continues its march around the sun.... It is by His power that vegetation is caused to flourish, that every leaf appears and every flower blooms.... In God we live and move and have our being" (*Testimonies for the Church*, vol. 8, pp. 259, 260).

I did not know what to do with the notion of the LORD being continually involved with the routine operation of nature. In an essay, published in 2000, I wrote: "These and other similar passages in the writings of Ellen White suggest the LORD's intimate engagement in the operation of our world. But science and scientists are clueless to deal with such a concept. To us matter behaves in a perfectly predictable manner, obeying the fundamental laws of gravity, attractions between positive and negative charges, etc. Chemical properties of each element depend on the configuration of its valence electrons. Biochemical properties of living matter are understood, based on the characteristics of proteins, nucleic acids, carbohydrates, and lipids."

"While it may be suggested that the Lord works precisely through these and other laws of nature, it is a very unsatisfactory solution because it is not testable. Moreover, it renders the Lord directly responsible for every undesirable physical event in the world. If the Lord directly pushes atoms and molecules around, then He would surely stop doing it when it comes to an explosion by a suicide bomber!

"Accepting the Creatorship of the Lord does imply that all matter proceeded from Him, and that the Lord is aware of every atom in the Universe. But it does not necessarily follow that the Lord micromanages the Universe through actively superintending every chemical change.

"I am more comfortable letting the mystery of the nature of the Lord's involvement with our world linger until we enroll in a university on the earth made new. There I will be perfectly willing to revise the organizational chart [of Figure 2] to include the vital connections between humanity and the rest of the Universe" (George T. Javor, *Biblical Approaches to Biology. Christ in the Classroom: Adventist Approaches to the Integration of Faith and Learning*, Vol. 26A, Humberto M. Rasi, compiler, 2000, pp. 481–502).

This comment was intended for my colleagues in the Seventh-day Adventist scientific community, cautioning them that we should not be too smug about our understanding of the relationship between the Creator and His creation. But I did not offer any additional insight into this problem.

Thirteen years passed without anyone advancing this topic. In 2013, an opportunity came to speak at a conference on teaching science in Seventh-day Adventist schools, organized by the Geoscience Institute. Preparing the material that is presented in the first part of this article, it occurred to me that the LORD's involvement in nature may be much more robust than I previously imagined.

What if all the laws of physics, chemistry, and biochemistry, which govern the behavior of matter, were dependent on the continual expression of the Creator's power? This power would be required for the continual existence of subatomic particles, for the phenomenon of gravity, magnetic forces, positive and negative charges, etc.

This view of reality affirms that nothing in the created universe is truly independent of the Creator. All created entities, animate and inanimate, owe their origins and continued existence, nanosecond by nanosecond, to the maintenance of the Creator. This power underwrites all the known laws that science has identified. The LORD does not micromanage all chemical reactions in nature. Rather, He provides a continuous sustaining power, without which nothing could exist.

The Creator and his creation are separate entities, but nothing exists without being sustained by the Creator, as described in Hebrews 1:1–3.

> In the past God spoke to our ancestors through the prophets at many times and in various ways, but in these last days he has spoken to us by his Son, whom he appointed heir of all things, and through whom also he made the universe. The Son is the radiance

of God's glory and the exact representation of his being, sustaining all things by his powerful word. After he had provided purification for sins, he sat down at the right hand of the Majesty in heaven. (Heb. 1:1–3, NIV)

I call this *radical creationism*. This concept emphasizes the reality of the Creator's continued sustenance of creation. It removes the possibility that matter on its own would organize itself into planets, stars, and galaxies in space. It negates the chemical evolutionary doctrine that life arose spontaneously on a hypothetical primordial earth, and it eliminates any notion of the evolutionary "tree of life," which purports to show the evolutionary linkages of all organisms.

Radical creationism moves the doctrine of creation from the past to the present, in that we continually are mindful of the Creator's sustaining power in our lives. Sabbath keeping is not just about the past, but also about the present and the future. We are safe from destruction in the hands of our caring Creator!

To be sure, the LORD "maketh his sun to rise on the evil and on the good, and sendeth rain on the just and on the unjust" (Matt. 5:45). The LORD underwrites the existence of evil with the hope that some will turn from their destructive ways. This fact alone ensures that current conditions will not last indefinitely. The tide of evil washing over the world must pain the LORD greatly.

The many-faceted implications of radical creationism remain to be identified. The immediate task for Christian teachers of biology is to teach their students not to view organisms as sophisticated machines but rather as precious expressions of the Creator's sustaining love.

For further reading, see:
George T. Javor, *Evidences for Creation*, Review and Herald Publishing Association, Hagerstown, MD, 2005.
George T. Javor, *A Scientist Celebrates Creation*, TEACH Services, Inc., Ringgold, GA, 2012.

CHAPTER 8

Searching for the Creator Through a Study of a Bacterium

(This chapter was included in the series "Online Classic Apologetics," in which the term "classic" is used to mean "of the highest order.")

"Until a man has found God and been found by God, he begins at no beginning, he works to no end. He may have his friendships, his partial loyalties, his scraps of honor. But all these things fall into place, and life falls into place, only with God."—H. G. Wells.

The heart of every human being, whether aware of it or not, continually searches for God. The search takes many forms and goes in many directions. Happy are those who recognize the nature of their restlessness and orient their search toward God. The answer to the question "Canst thou by searching find out God?" (Job 11:7), is, "Yes—through God's grace" (Deut. 4:29). And finding God is not the end but the beginning of a fulfilled life.

Finding God is a very personal matter that cannot be accomplished for someone else. But the fruits of one's search may be shared. Seeking for God may begin with Bible study, but it does not have to end there. The signature of the LORD is unobtrusively present in many places—like an artist's autograph on a painting. The object of this essay is to study one such signature.

Underlying this exercise is the conviction that we are endowed with the capacity both to comprehend the physical realities of our world and to seek for deep meaning in them. The most profound understanding that may be culled from the study of nature is the gaining of insights into the character of the Maker of the universe.

Of one of the wisest men, the following was written: "Solomon took an especial interest in natural history, but his researches were not confined to any one branch of learning. Through a diligent study of all created things, both animate and inanimate, he gained a clear conception of the Creator. In the forces of nature, in the mineral and the animal world, and in every tree and shrub and flower, he saw a revelation of God's wisdom; and as he sought to learn more and more, his knowledge of God and his love for Him constantly increased" (*Prophets and Kings*, 1917, p. 33).

In this essay, it will be a given that the universe and our world are God's creation, as described in the Bible. I state this at the outset because today's academia is largely in the evolutionist camp. As a scientist, I frequently find myself taking a polemic stance in defense of creationism. However, in doing so, I can easily lose sight of nature as a revealer of its Creator.

"In the beginning God created the heaven and the earth" (Gen. 1:1). This verse is likely to refer to the creation of the earth and of its immediate surrounding support system, including possibly the solar system. Ours is not the first world, and we are not the first beings created. When Earth was born, "the sons of God shouted for joy" (Job 38:7). As astronomers began examining our cosmic neighborhood, they discovered that the solar system is part of a galaxy of perhaps as many as 100 billion stars, each at least the size of our own sun.

The Milky Way, in turn, is part of a group of galaxies called the "local group." It now appears that the universe contains maybe a billion galaxies, which are arranged in strings, clusters, and superclusters (Yakov B. Zeldovich, Jaan Einasto, and Sergei F. Shandarin, "Giant Voids in the Universe," *Nature*, Vol. 300 [2 Dec. 1982], p. 407). The Creator of the heaven and the earth of Genesis 1:1 is also the Creator of these unnumbered galaxies.

There was a time, back in the distant past, when there was no universe, but only the triune Godhead. The material universe came into existence as the result of God's creative action. The Creator was not dependent on preexisting matter for creation: "For he spake, and it was done; he commanded, and it stood fast" (Ps. 33:9).

Matter as we know it is a stable form of energy. All the matter we now find in the universe had to come from God. Einstein's famous equation, $E = mc^2$, permits us to estimate the energy cost of creating 1.8×10^{53} grams of matter, which is an estimate of the total visible mass of the universe. The energy consumed was approximately 1.6×10^{63} joules. The total mass of the universe is approximated by assuming that the Earth's mass is 6×10^{27} grams, our sun's mass is 3×10^3 Earth's mass, the Milky Way is 1×10^{11} the sun's mass, and the universe contains 1×10^9 galaxies that are the same size as the Milky Way. Einstein's equation of $E = mc^2$ was used to calculate the energy equivalent of the universe's estimated matter ($c = 3 \times 10^3$ km/sec; the results for energy were obtained by meters x Newton's = joules). This amount of energy could supply the earth's energy needs for approximately 10^{45} years (S. Fred Singer, in "Human Energy Production as a Process in the Biosphere," *Scientific American*, September 1970, p. 175). Here the energy consumption was expressed in BTUs (British thermal units). These were converted to joules, with a BTU equaling 1,055 joules.

Why God created the universe will be the subject of contemplation for a long time. The amount of energy invested to produce just the raw materials needed for the universe is utterly beyond our comprehension, though we can appreciate that it was prodigious. One cannot help wondering about the connection between the living God and inanimate matter, which came forth from Him. While we would not want to say that God is "in" matter (as the pantheist does), we affirm that all matter belongs to God by virtue of having proceeded from Him. It is a reasonable postulate that the LORD has absolute control over the inanimate world. Perhaps He keeps track of every atom. "The very hairs of your head are all numbered" (Matt. 10:30). From such a perspective, it does not seem difficult to understand how Jesus could multiply the loaves and fishes, calm the Sea of Galilee, or command Lazarus to walk out of his grave.

Relying on our eyes, we notice objects in our environment ranging in size from 0.0001 centimeters to approximately 10,000 centimeters in size, or eight orders of magnitude. With the help of the microscope and the telescope, we become aware that we are part of a much larger universe, spanning perhaps 40 orders of magnitude. (The universe's diameter is about 93 billion light-years, and a light-year is about 9.46 trillion kilometers.) The Creator is the Designer and the Caretaker of that which is, from our perspective, very large and very small. We understand some things about the very small, but the reality of the very large is puzzling. Existence was cozier in the universe of the Middle Ages, which was conceived as revolving around a flat Earth. What are those billions of galaxies doing out there?

One is tempted to come up with an outrageously egocentric theory to explain this oversupply of galaxies. Every galaxy could represent a different order of existence. Perhaps the Creator experimented over eons of time with creation, not being completely satisfied until He came to the creation of the Earth. Here He created beings in God's image, an effort with which the Creator was finally satisfied. So He pronounced it "very good."

Of course, we don't have any information regarding why there are so many galaxies. The Bible text, "For as the heavens are higher than the earth, so are my ways higher than your ways, and my thoughts than your thoughts" (Isa. 55:9), comes to mind.

As a youngster in Hungary I wanted to become a doctor; but, after coming to the United States, I studied chemistry in college. I was particularly attracted to organic chemistry. In my junior year, I took a course in biochemistry, and I realized that this subject combined my interests of chemistry and medicine. So I continued studying biochemistry in graduate school.

Biochemistry is the study of the chemistry of life. Paradoxically, to study the chemistry of living matter, one must take the living tissue apart, thereby killing it. When I was a graduate student, one of my laboratory exercises called for taking apart a white albino rat, removing its liver and studying its cholesterol metabolism. I performed the experiment as required, heart-sickened from the act of killing the cute laboratory rat.

I realized that as a biochemist I may have to do this many times during my career, and I wasn't sure that I wanted to do that. Fortunately, I discovered that not all biochemists work with animal tissues. There were professors in my department who worked with bacteria. I arranged to do my thesis work in such a laboratory, and I have continued to work with bacterial cells ever since. To be sure, studying the biochemistry of bacteria also involves killing bacterial cells. However, bacteria belong to a different class of living entities than albino rats. Every time we use mouthwash we are guilty of killing microorganisms in the oral cavity.

It so happens that the greatest advances in biochemistry in the past fifty years have come from studying the biochemistry of bacteria. Among the many species of bacteria, one organism towers above all others in significance as the best-studied model organism: *Escherichia coli*. Some researchers have jokingly gone so far as to classify all living organisms into just two categories—the coli and the "un-coli"!

It is now clear that there are overarching similarities in all biological matter, from bacteria to human beings. These similarities surface when one compares the gross chemical composition of various organisms or the biochemical logic that animates them.

All living matter is composed of cells. Some organisms, such as humans, are organized from millions of different cells. We have skin, muscle, bone, liver, and brain cells. All are different in structure and function yet retain certain similarities with each other. Other organisms may contain fewer cells or consist of only a single self-contained cell. The most fundamental unit of life is the cell. When a cell is taken apart, life disappears. Consequently, you can appreciate how the in-depth study of *E. coli* could lead to advances in our general knowledge of life.

Escherichia coli is a rod-shaped microorganism that was first isolated early in the past century from the stool of a convalescent diphtheria patient by Theodor Escherich, a German physician. It is part of the normal flora of many species of microorganisms, residing in the large intestines of vertebrates. Because of its relative non-virulence and ease of cultivation in the laboratory, it became the organism of choice for scientific experimenters. After nearly a century of research, more is known about *E. coli* than just about any other life form. The study of *E. coli* informs us about the logic of life in general. One of my teachers in graduate school put it this way: "An elephant is like an *E. coli*, only more so."

Of course, this is a major oversimplification. Bacteria, which are unicellular, are life forms markedly different from the considerably more complex multicellular organisms. Moreover, there are considerable variations even among the thousands of different types of microorganisms.

Nevertheless, the rapid advances in biochemistry over the past fifty years can be partly attributed to the diligent work of thousands of scientists on *E. coli*. In 1996, a two-volume set of articles was published, consisting of 155 chapters on 2,800 oversized pages summarizing some of our knowledge of this organism (Frederick C. Neidhardt, ed., *Escherichia coli and Salmonella: Cellular and Molecular Biology*, 1996). This was the second edition of a similar effort that had appeared in 1987. In the second edition, essentially all the articles from the first edition had to be rewritten because of the rapid accumulation of knowledge. Six years later, with the contents of the second volume once again hopelessly outdated, the editors took a different tack. They constructed a website, accessible by subscription, where the material is continuously updated. It is apparent that even this comparatively simple life form is a very complex organism.

Humanity existed for more than five millennia without the slightest inkling of microorganisms. In the instructions to the Israelites for the practice of personal hygiene and for the isolation of lepers, we now recognize effective preventive measures against the spreading of contagious diseases. Had these instructions been followed in the Middle Ages, humanity would have been spared many epidemics.

Microorganisms such as bacteria are better known by the public as "germs," and germs definitely have an image problem. Why did the good Creator make bacteria? The short answer is that life on Planet Earth would be impossible without them. The most vital role they play is the conversion of nitrogen gas in the air to useful nitrates in the soil, which plants need for growth. In addition, some bacteria participate in the capture of solar energy through photosynthesis, and others implement the biodegradation of dead organic matter.

Microorganisms cover our skin and grow in our oral cavities and in our intestines; they protect us from harmful biological agents. *E. coli* bacteria help create an oxygen-free environment in our colon for the benefit of necessary anaerobic organisms, which aid digestion. They also secrete, for our use, a water-soluble B vitamin, choline. Germ-free laboratory animals are much more vulnerable to infection and disease than their germ-carrying counterparts.

The biosphere is essentially coated with microorganisms. It is projected that nowhere can a gram of soil be found on the surface of the Earth, including in the Sahara, that contains less than 10,000 microorganisms. Bacteria belong to the robotic class of living matter created by the LORD. Other living robots include trees, plants, and flowers.

Bio-robots are living organisms without nervous systems. Biologists place them in the vegetable kingdom. These organisms, although they respond to appropriate external stimuli, are unaware of their existence. They perform photosynthesis and are the ultimate sources of food on Earth. Microorganisms belong to this kingdom as well.

The vast majority of microorganisms in our environment are harmless except if they find their way accidentally (such as during surgery) into a nutritionally rich environment within the body. Most disease-causing microorganisms, including the pathogenic variety of *E. coli*, have acquired extra pieces of genetic material called "plasmids." These extra chromosomal pieces contain the disease causing genes.

Among the most dangerous *E. coli* is the strain 0157:H7, which causes hemorrhagic colitis. This organism colonizes in the small intestine of the host and secretes large quantities of toxins, which damage the lining of the intestine. By acquiring a plasmid with genetic information for these toxins somewhere in the past, a harmless strain of *E. coli* was converted to a dangerous pathogen. Just where these harmful plasmids came from no one knows. But the Christian suspects that "an enemy hath done this" (Matt. 13:28).

Like all living matter, bacteria are seventy percent water and nearly thirty percent biopolymer. The remarkable similarity—but not identity—in composition and internal workings among all living entities is a fruitful theme for contemplation. Conceptually all organisms may be thought of as variations on one or more themes. *E. coli* is one of about 5,000 different types of bacteria. Each organism carries within itself a set of biological agents called "restriction enzymes." These enzymes are programmed to recognize the host's own genetic material and, if the need arises, destroy any foreign genetic substance. These agents guard the genetic uniqueness of each type of microorganism,

Similarities among living entities ensure that their needs are similar or at least compatible with each other, simplifying the task of supplying nutrients to all. A very large portion of the biosphere utilizes the products of photosynthesis in the form of plant products. Researchers have also observed that the discarded waste of one organism becomes a useful resource for another. The best-known example is the production of carbon dioxide by non-photosynthetic entities, to the delight of photosynthetic organisms. The resources available for all organisms exquisitely match their needs. One hears echoes of Philippians 4:19: "My God shall supply all your need."

Biopolymers are complex substances, composed of thousands (or in the case of RNA and DNA, millions) of atoms of carbon, hydrogen, oxygen, nitrogen, phosphorous, and sulfur. The four classes of biopolymers are proteins, nucleic acids, polysaccharides, and lipids. An *E. coli* cell contains millions of polymer molecules. Sonic polymers are needed in many copies; others, in just a few. It is estimated that every cell uses approximately 3,000 different kinds of polymers, each with a different structure and a different function.

Bacterial cells have to accomplish four major tasks for their existence: (1) harness energy from the environment and utilize it for biosynthesis and growth, (2) manufacture the building blocks of its biopolymers, (3) synthesize biopolymers, and (4) break down existing biopolymers for the purpose of continuous renewal.

Energy is the driving force behind all life. Almost everything connected with life, chemically speaking, is an uphill event. Energy is required for the transport of all nutrients into the cell, for the manufacture of biopolymers and subcellular structures, and for the maintenance of the physical integrity of the cell. *E. coli* is able to extract chemical energy from such organic substances as glucose. Glucose, a product of plant photosynthesis, contains some of the sun's light energy, which was captured by the plant. Thus, an *E. coli* bacterium, just like the rest of us, runs on solar energy.

The solar energy trapped in glucose may be released by non-biological means. One can burn glucose, for example, to release 4 kilocalories/gram of energy as heat. But such a burst of heat energy is useless to living matter. In the cell, glucose is utilized by an ingenious process equivalent to a very slow burning called "glycolysis" and by the citric acid cycle. These biochemical pathways methodically dismantle the sugar molecule to smaller and smaller substances, which contain less and less energy.

The energy is drained out of glucose in the form of high-energy electrons that are captured by designated electron-carrier substances. The electron carriers transport the electrons into a collecting device called "the respiratory chain." Here the electrons flow through the cell's membranes as tiny micro currents of electricity. This process transfers the electrons' energy into a charge difference on the two sides of the cell membrane. The energy of the charge separation, in turn, pushes the synthesis of ATP molecules to completion. ATP—adenosine triphosphate—is the chief carrier of chemical energy in the cell.

Thus, one can trace the flow of energy from the sun to the ATP molecules. We feel pretty smug when we are able to operate a device needing 9-volt direct current by plugging it into a 110-volt alternating current socket through an appropriate adapter. In nature, we see the energy of the sun's nuclear reactor, which generates millions of degrees of heat, being transformed to operate bacterial and other metabolisms and running at a fraction of a volt of electricity. Here we observe the Creator smoothly connecting the very large with the very small.

Chemical factories are designed to produce quantities of substances for sale. Other enterprises purchase their products and use them for manufacture or for research. Each bacterial cell is a miniature chemical factory, where the manufactured products are used in-house to create more chemical factories. In fact, even during the manufacturing process, the factory is expanding and continually using up products as they come off the assembly lines.

> *Compared to the bacterial factory, chemical factories have only modest inventories of products.*

Compared to the bacterial factory, chemical factories have only modest inventories of products. When we consider that bacteria manufacture all their biochemical intermediates, polymer-building subunits, and polymers, the number of manufactured products is in the neighborhood of 3,000 different substances. Each of these products has a function in the cell. As soon as they are produced, the products are integrated into their proper places. The assembly lines are also regulated, so that no resources are wasted. The regulation includes an immediate acceleration or slowing down of the manufacturing rate, depending on the needs of the cell, as well as an adjustment in the long-term production of the "factory equipment."

These factories have no director, foremen, or assembly workers; they are 100 percent automated.

If someone would write a fictional narrative about self-replicating, fully automated, microscopic factories, it would probably be rejected out of hand as completely absurd. But they actually exist! When we wonder about the marvels of the future new earth, we should remember that we live in a world filled to the brim with miraculous manifestations. Scientists are busy trying to understand the molecular explanations of some of these marvels. Those who dig the deepest into the mysteries of nature are usually the most aware of how little we actually know or understand.

The genetic material (DNA) of *E. coli* consists of 4.6 million pairs of nucleotides. It contains data for the correct structure of every protein of the bacterium, and for regulating their production. Indirectly, through the action of proteins, every aspect of the metabolism and the infrastructure of the organism are coded into its genome. In terms of information density, estimates are that one cubic micrometer of DNA contains 150 megabytes of information. This density is several orders of magnitude greater than the optical storage capacity of a currently available digital video disk (DVD). The complete genetic information of *E. coli* takes up 1,100 pages in standard book form.

A surprising discovery was made in the 1950s: bacteria biopolymers (and the biopolymers in all other organisms), which represent the investment of a great deal of energy, are periodically dismantled and replaced by new polymers. In time, researchers realized that, during normal metabolism, polymers are damaged and lose their functions. Therefore, researchers realized that, as a form of "preventive maintenance," the turnover process is essential for all living matter. The Creator is seen here as anticipating problems and instituting steps for their prevention.

There is evidence to indicate that *E. coli* was created to live in our colon. The bacterium has three genes that enable it to utilize the "double-sugar" lactose, sometimes called "milk sugar." These genes direct the synthesis of three proteins. One of the proteins carries lactose into the cell. A second protein cleaves lactose into two "single" sugars—glucose and galactose. The third protein converts unusable potentially galactose-related sugars for secretion from the cell. Because lactose does not usually reach the colon of humans (it is usually digested and absorbed before reaching the colon), the three "lactose genes" are ordinarily silent. In fact, maybe half to two thirds of *E. coli*'s genes are routinely silent, waiting to be activated when needed. Apparently a great deal of metabolic flexibility was programmed into this organism.

Until recently, the physiological role of the lactose genes in *E. coli* was something of a mystery. Then it was discovered that an indigestible plant product, galactosyl-glycerol, activates the lactose genes extremely well. It is now understood that *E. coli* was designed to take advantage of the plant diet of its host.

There is also evidence that we were designed to have *E. coli* inhabiting our colons. Alcohol dehydrogenase, an enzyme in our liver, promotes the detoxification of alcohol in our blood. But why should we have such an enzyme in our liver when alcohol is not produced by any of our body's metabolic reactions? We now know that *E. coli* cells in our colon in fact produce a small amount of alcohol as one of their normal fermentation products. The alcohol thus produced undoubtedly would make us tipsy continuously were it not for our liver's alcohol dehydrogenase, which nicely neutralizes it.

In Isaiah 45:18 we read that the LORD created the earth to be inhabited. But He could have placed humanity in a sterile, non-living environment. Instead, God immersed humanity into a sea of life and gave human beings the task of managing the biosphere. Did He assign this task because it would have been too much for Him to do? Not likely. Rather, Adam was asked to provide names for the created organisms, thereby becoming a co-worker with God. One wonders if the LORD told Adam about *E. coli* in his colon or if this was to have been revealed to him at a later time.

Summary

It is a privilege beyond words to be alive in this amazing world. We are surrounded with innumerable miraculous phenomena. Our very act of probing is amazing. We, a collection of inanimate atoms and molecules, are put together in such a way as to have the ability and curiosity to probe and judge.

We cannot overstate or exaggerate the greatness of God. The scope of this essay does not permit the consideration of even a small fraction of the amazing features uncovered in the study of the lowly microscopic bio-robot *E. coli*. We did not explore its capacity to sense favorable nutritional gradients or chemical insults from the environment. Nor did we recount its documented nutritional versatility or its capacity to withstand mechanical stresses, which amount to thousands of times the force of gravity.

But from a small selection of examples, it is clear that even at this microscopic level the Creator was fully able to implement a variation on the theme of life, completely appropriate for its intended functions. Although we now "see through a glass, darkly" (1 Cor. 13:12), through a diligent study of His works, we will continue to draw closer to our Maker throughout eternity.

CHAPTER 9
The Mystery of Life

The study of living matter is at the center of all current scientific efforts. Recent triumphs include the cloning of Dolly the sheep and the acquisition of the complete sequence of three billion nucleotides of the human genome (Eric S. Lander et al. at the Whitehead Institute for Biomedical Research, Cambridge, MA, "Initial sequencing and analysis of the human genome," *Nature*, Vol. 409, Iss. 6822 [15 Feb. 2001], pp. 860–921; see also J. Craig Vent et al., "The sequence of the human genome," *Science*, Vol. 291, Iss. 5507 [16 Feb. 2001], pp. 1304–1351). But strangely, life itself is not the object of much study. Scientists seem to take the existence of life for granted. It is difficult to find any extended discussion on the essence of life in currently available monographs or textbooks. These publications explain well how living matter is put together and how its components function. But such information is not enough to explain life because the constituents of living matter themselves are lifeless.

Suppose we take apart the living matter and then recombine the isolated components. The work will yield an impressive collection of inert substances—but not life. So far, science has not created living matter in the laboratory. Is this because living matter contains one or more components that cannot be supplied by the chemist? The answer, as developed in this article, will suggest an important point regarding the origin of life.

More than 100 years ago Louis Pasteur and others proved the folly of abiogenesis—the spontaneous transformation of non-living matter into living organisms. Biologists now simply say, "Life comes only from life." Nevertheless, scientists generally accept the concept that life developed abiologically on a primordial Earth. In doing so, they conveniently assert that conditions on a "primordial world" were conducive to generate life spontaneously.

Others theorize that perhaps life was imported to Earth from outer space. But while Earth is covered with millions of different species of organisms, there is no evidence of life anywhere else in the solar system. And beyond it, there is 4.37 light-years of empty space until the nearest star, Alpha Centauri.

The last logical option for the origin of life is creation by a supernatural Creator. But scientists, in their attempt to explain everything by natural laws, reject the creation option as being outside the scientific realm.

Life is not a tangible entity. It cannot be put into a jar or handled. We only see "life" in association with unique kinds of matter that have the capacity to grow and divide into replicas, can respond to various external stimuli, and utilize light or chemical energy to accomplish all these things. Such an analysis of life may seem too materialistic to many who perceive that the Bible teaches a different view of life—one which does not insist that it be associated with matter. While there may well exist larger realities of life inaccessible to us, so far as science is concerned, we experience life on Earth only in association with matter. The Bible supports the notion that life as we know it on Earth is associated with matter. Says the book of Genesis: "The LORD God formed man of the dust of the ground, and breathed into his nostrils the breath of life; and man became a living soul" (Gen. 2:7). A combination of the breath of life and the dust of the ground gave rise to the living person. Similarly, a person dies when "his breath goeth forth, he returneth to his earth; in that very day his thoughts perish" (Ps. 146:4). The "return to earth" marks the end point of human existence. While one can speculate on the meaning of the "breath of life" and of the person's "breath," it is clear that life as experienced on Earth does not continue after death. The Bible does not mention anything about a disembodied form of life. To embrace the material basis of life on Earth, therefore, does not make one a materialist.

The term "life" has different meanings, depending on whether it refers to an organism, an organ, or a cell. Human organs may continue to live after a person's death if, within a certain time, the organs are transplanted into another living person. Survival of a transplanted liver, kidney, or heart means something quite different from human "life." Furthermore, the life of each organ depends on the vitality of its cells.

All manifestations of life depend on living cells, the most fundamental units of living matter. When a live cell is taken apart, a collection of very complex but lifeless subcellular structures remain—membranes, nuclei, mitochondria, ribosomes, etc.

Is there an unbroken continuum between living and non-living matter, as some would assert? If there is, the question of the origin of life becomes moot. Moving from one state to the other would be similar to other chemical transformations. Examples of organisms that supposedly bridge the chasm between the living and non-living include viruses, prions, mycoplasmas, rickettsiae, and clamidiae.

In fact, viruses and prions are biologically active but non-living entities. The term "live virus" is a misnomer, even though a virus is a biologically active agent that infects living cells. Prions are unique proteins that have the capacity to alter the structures of other proteins (Stanley B. Prusiner, "Prion Diseases and the BSE Crisis," *Science*, Vol. 278, Iss. 5336 [10 Oct. 1997], pp. 245–251). The newly changed proteins, in turn, acquire prion-type activity, creating a domino effect of protein alteration. This property of prions renders them infectious. For reproduction, prions, like viruses, need living cells.

Rickettsiae, chlamidiae, and mycoplasmas, on the other hand, are among the smallest known living organisms. The first two have serious metabolic deficiencies and can only exist as obligate intracellular parasites. There is a wide gap between living and non-living matter.

This is best reflected in our inability to bring non-living matter to life in the laboratory.

Structurally, living matter is composed of a combination of water and large fragile, lifeless molecules, proteins, polysaccharides, nucleic acids, and lipids. Table 1 lists the gross chemical composition of a typical bacterial cell, *Escherichia coli*.

Table 1 **Components of *Escherichia coli* Cells**			
Components	Percent of total weigh	Number of molecules per cell	Number of different kinds of molecules
Water	70	24.3 billion	1
Proteins	15	2.4 million	4,000 approx.
Nucleic acids	7	255 thousand	600
Polysaccharides	3	1.4 million	3
Lipids	2	22 million	50–100
Metabolic intermediates	2	many millions	800
Minerals	1	many millions	10–30

Water serves as the medium in which all chemical changes occur. Proteins and lipids are the principal structural components of cells. Proteins also control all chemical changes. Without chemical changes, life cannot exist. How proteins interact with chemical changes is central to understanding the chemical basis of life.

Proteins come in thousands of different forms, each with unique chemical and physical properties. This diversity is due to their size—each protein can contain hundreds of amino acids, and there are twenty different amino acids. What each protein is capable of doing depends on the order in which its amino acids are linked. To understand this feature of biology, consider the analogy of written language.

In any language, the meaning of words depends on the sequences of letters. In English, for example, we have twenty-six letters. Out of these we make words. An estimated 500,000 different combinations of letters are recognized as meaningful words. With some effort, we could produce another 500,000 or more nonsensical combinations. Similarly, the millions of different proteins represent but a tiny fraction of all possible combinations of amino acids. (The number of possible different sequences for a 100 amino acid-long protein is 1.2×100^{130}, which is 12 followed by 129 zeros!)

When words are misspelled, their meaning is garbled or lost. In similar fashion, for proteins to function properly, their amino acids must follow one another in the correct order. The results of alterations in the amino acid sequence can be drastic. Hemoglobin, the oxygen-carrying protein in blood, is built from four chains of more than 140 amino acids each. In sickle cell anemia, an inherited disease, an altered amino acid occurs in the sixth position of a specific sequence of 146. This change causes distortion of the red blood cells, resulting in anemia and many other problems.

How does the protein-building apparatus know the correct amino acid sequences for each of the thousands of proteins? The chromosomes of each cell are libraries filled with just such information. Each volume in this library is a gene. When the cell needs a particular protein, it activates the protein's gene, and synthesis begins. The details of this process can be found in any current biology or biochemistry textbook. Here it is sufficient to note that more than 100 separate chemical events must occur for protein synthesis to happen.

All manifestations of life depend on chemical changes. These changes happen when atomic clusters (molecules) gain, lose, or rearrange atoms. A class of proteins, known as enzymes, binds specific molecules and facilitates their chemical transformations. In *E. coli*, there are about 3,000 different types of enzymes, facilitating 3,000 different chemical changes.

Enzymes speed up reactions enormously. This could be a huge problem because, once the reaction is completed, its end point—known as equilibrium—is reached, and no further chemical changes occur. Because life depends on chemical changes, when all reactions reach their end points, the cell dies.

Amazingly, in living matter *none of the reactions ever reach equilibrium.* This is so because the chemical transformations are *interlinked* so that the product of one chemical change forms the starting substance of the next. If biological molecules were represented by capital letters of the alphabet, a typical sequence of chemical conversions would look like Figure 3 below.

Figure 3				
enzyme 1	enzyme 2	enzyme 3	enzyme 4	enzyme 5
A ⟶	B ⟶	C ⟶	D ⟶	E ⟶ F

Such a sequence, or "biochemical pathway," resembles a factory assembly line. The end product of this particular pathway, substance F, is utilized by the cell. Therefore, it does not accumulate. In living matter, every one of the millions of molecules (Table 1) is inventoried. Any shortage or excess immediately results in adjustment in the rates of chemical transformations.

Figure 4 below shows that, in a live cell, matter is organized into successively more complex hierarchies. The arrows represent biochemical pathways, leading from simple to complex substances. The interdependence among cellular components in the vertical direction parallels the logical relationships of written language among letters, words, and sentences all the way to the level of a book.

However, the degree of tolerance for errors is much smaller in biology. Misspelled words, garbled sentences, or missing paragraphs may *not* render a document useless. But, given the tight functional interdependence of its components, cells would be in big trouble if they lack a full complement of parts.

There is horizontal complementation among cell components as well. For example, proteins cannot be manufactured without assistance from nucleic acids, and nucleic acids cannot be made without proteins. From a chemical evolutionary perspective, this problem resembles the classic "Which came first, the chicken or the egg?" (See Figure 4.)

Each biosynthetic pathway feeds into successively more complex levels of organization of matter. Every pathway is regulated so that its output is appropriate for the needs of the cell. The life of the cell depends on the harmonious and nearly simultaneous operation of its many components. During balanced growth, a steady-state exists; that is, there are only minimal perturbations in the flux of matter through the pathways. Since none of the reactions is permitted to reach its end point, *each of the thousands of interlinked chemical reactions is in a non-equilibrium steady-state.*

If there are forces in nature that bring about life, we should search diligently to discover and harness them. If abiogenesis is possible, it could be harnessed to restore dead cells, organs, and even organisms to life. Who

Figure 4
Organization of Matter in the Cell

Level number	Components				An analogy
1. Precursors	Carbon dioxide, water, ammonia				1. Letters
↓	↓ ↗			↗	↓
2. Building blocks	Amino acids	Monosaccharides	Nucleotides	Fatty acids + glycerol	2. Words
↓	↓	↓	↓	↓	↓
3. Polymers	Proteins	Polysaccharides	Nucleic acids	Lipids	3. Sentences
↓	↘	↓	↓	↙	↓
4. Supramolecular assemblies	Enzyme complexes, ribosomes, etc.				4. Paragraphs
↓	↓		→		↓
5. Organelles	Membranes, nuclei, mitochondria, etc.				5. Chapters
↓	↓				↓
6. Cell	Cell				6. Book

would argue that creating living matter, or reversing death, would not be humanity's most significant scientific achievement?

However, the history of biochemistry suggests that this possibility is unlikely. In the 1920s, when Oparin and Haldane first proposed that life originated spontaneously on a primordial Earth, biochemistry was in its infancy. The concept itself was an elaboration of Darwin's idea that life arose in some "warm little pond" (Charles Darwin's 1871 letter, quoted by Francis Darwin, *The Life and Letters of Charles Darwin*, vol. 2, 1887, pp. 202, 203, footnote). The first metabolic pathway was described only in the 1930s. The structure and function of the genetic material began to be understood in the 1950s. The first amino acid sequence of a protein, insulin, was mapped in 1955, and the first nucleotide sequence of the chromosome of a living organism was not published until 1995.

As the chemical basis of life began to be understood better, it turned out to be far more complex than originally imagined, and the early abiogenetic suggestions should have been reconsidered. Instead, science embarked on a half-a-century-long journey to demonstrate experimentally the plausibility of spontaneous abiogenesis.

The first experiments suggesting the plausibility of chemical evolution were performed by Stanley Miller, who in 1953 reported the synthesis of amino acids and other organic substances under simulated primordial conditions (Stanley L. Miller, "A Production of Amino Acids Under Possible Primitive Earth Conditions," *Science*, Vol. 117, Iss. 3046 [15 May 1953], pp. 528, 529). Subsequently, a subdiscipline emerged, which provided laboratory evidence of the production of nineteen or twenty amino acids and four or five nitrogenous bases needed for nucleic acid synthesis, monosaccharaides, and fatty acids—all under varying hypothetical primordial conditions (Charles B. Thaxton, Walter L. Bradley, and Roger L. Olsen, *The Mystery of Life's Origins*, 1984, p. 38). All these substances are the components from which the large biopolymers are made, projecting the possibility of the primordial production of biopolymers.

However, actually demonstrating the linking of building blocks into chains of polymers could not be accomplished. Every link between building-block type substances requires the removal of water. This is next to impossible in the aqueous environment of the hypothetical primordial oceans. Furthermore, the sequences in which amino acids are strung together in proteins, or nucleotides in nucleic acids, are what determine the function of these biopolymers. Outside of living matter, there are no known mechanisms to ensure meaningful, reproducible sequences in proteins or nucleic acids.

Under simulated primordial conditions, protein-like matter has been made by heating powders of amino acids to high temperatures. However, these "proteinoids" were amino acids randomly linked by unnatural bonds that have little resemblance to actual proteins (Sidney W. Fox and Klaus Dose, *Molecular Evolution and the Origins of Life*, 1977, second edition).

Nucleotides, the building blocks of nucleic acids, have not yet been synthesized under simulated primordial conditions. This is a formidable task, which requires attaching a purine or pyrimidine base to a sugar and that to a phosphate. The challenge here is not only the removal of water but that these three components may be linked together in dozens of different ways. All combinations but one are biologically irrelevant. Needless to say, nucleic acids have not been synthesized.

But this has not stopped many scientists from postulating that the earliest living cells contained primarily ribonucleic acids. This "RNA world" hypothesis gained popularity after it was discovered that certain RNA molecules had catalytic activities. Until then, it was believed that catalysis was the exclusive province of proteins.

Even though it is not possible to make biologically useful biopolymers under simulated primordial conditions, we can obtain them from once-living cells. Mixing these isolated biopolymers shortcuts chemical evolution, making it possible to test whether life will start from such a mixture. But in such preparations everything is at equilibrium. Since life happens only when all chemical events within the cell are in a state of non-equilibrium, the best that can be accomplished by this method is the assembly of dead cells.

We know exactly how to create living matter: first, design and synthesize a few thousand different molecular machines that are capable of converting simple substances, commonly available in the environment, into complex biopolymers. Second, make sure that such devices are capable of precise self-reproduction. Third, ensure that these units can sense their environment and adjust to any changes in it. Then it is only a matter of starting hundreds of biochemical pathways simultaneously, maintaining the non-equilibrium status of each chemical conversion by ensuring availability of a continuous supply of raw starting materials and providing for the efficient removal of waste substances. No problem, right? There's more.

A minimum requirement to create such complex biological devices is an absolute familiarity with matter on the atomic and molecular level. You will also need to have great ideas regarding the uses to be made of these complex living machines, hopefully in proportion with the effort expended in creating them. Fashioning living cells requires absolute control over every molecule, large and small. This is a capacity that science does not have. Chemists can manipulate large numbers of molecules from one form into another, but they cannot transport selected molecules across membranes to reverse conditions of equilibria. This is why we cannot reverse death.

So how did life originate on Earth? This article has revealed the great discrepancy between the biochemistry of living matter and of the claims of those who would explain its origins by spontaneous abiogenesis. Fifty years of biochemical research has shown unequivocally that, under any conditions, spontaneous abiogenesis is an impossibility. It is only a matter of time before the edifice called "chemical evolution" collapses under the weight of facts.

For the believer in the creation account of the Bible, the assertion that only the Creator can make life is not an argument for the "God of the gaps." We have a pretty good idea of what it takes to create life, only we cannot do it. It is an affirmation that life cannot exist apart from God. Indeed, life itself becomes an evidence for an all-wise Creator who chose to create life and share it with us.

CHAPTER 10

Creation in Focus

"There is not a great deal of interest in creation among Adventists" is what the manager of an Adventist Book Center told me. I was trying to persuade her to carry my recent book on creation, which was published by a self-supporting Seventh-day Adventist publishing house (see George T. Javor, *A Scientist Celebrates Creation*, TEACH Services, Inc., 2012). Frankly, her response shocked me. But, upon reflection, I realized that the manager simply shared her observation regarding the sales of books on creation.

Why would books on creation not sell? Perhaps it is because even though we are surrounded by evolutionary thinking in mainline media, academia, and even by some religious bodies, Seventh-day Adventists have long ago settled on the veracity of the biblical account of creation. After all, every seventh day we stop our busy lives to worship the Creator on the Sabbath. The very name of our denomination is a proclamation of our unshakable belief in creation!

The Seventh-day Adventist membership at large does not need to be convinced about the veracity of creation, hence the low level of interest in books on this topic. Of course, reality is a bit more complex. One just need to peruse the blogs on the websites of *Adventist Today* and *Spectrum* and listen to science teachers on Seventh-day Adventist campuses to find a large range of views on creation. We witnessed the recent controversy regarding the proper way to teach biology at La Sierra University.

Regardless of our interpretation of the creation account, for too long it has been merely one of the twenty-eight "fundamental beliefs." It has not received our full attention, and we are the poorer for it. The end-time call by the second angel of Revelation to worship the Creator is directed to us all (Rev. 14:7).

Long after evolution is forgotten, creation will dominate the thinking of the redeemed throughout eternity! Creation is the vehicle which draws us closer and closer to our Creator and Redeemer. The experience of the redeemed will be what Adam and Eve found in the Garden of Delight (that is, the Garden of Eden):

"The laws and operation of nature, which have engaged men's study for six thousand years, were opened to their minds by the Infinite Framer and Upholder of all.... On every leaf of the forest or stone of the mountains, in every shining star, in earth and air and sky, God's name was written. The order and harmony of creation spoke to them of infinite wisdom and power. They were ever discovering some attraction that filled their hearts with deeper love and called forth fresh expressions of gratitude. So long as they remained loyal to the divine law, their capacity to know, to enjoy, and to love would continue to increase. They would be constantly gaining new treasures of knowledge, discovering fresh springs of happiness, and obtaining clearer and yet clearer conceptions of the immeasurable unfailing love of God" (*Patriarchs and Prophets*, pp. 50, 51)

Even before sin entered the world, before the plan of salvation was revealed, the infinite wisdom, power, and immeasurable unfailing love of God were on display and accessible to our first parents as they studied the created world.

The Hebrew word for the creation of the world in Genesis 1:1 is *bara*. The word has no English equivalent. Once I heard a Hebrew scholar remark that only the Hebrew language has this particular term, which describes an act only God is capable of—"creating out of nothing." (It so happens, however, that my Hungarian mother tongue also has such a word, identical in meaning to *bara*. In Hungarian, the word is "teremt.")

Creating matter out of nothing violates the first law of thermodynamics, which states that the energy/matter content of the universe is constant. But the LORD is not confined to the three dimensions of the universe. In creating, the LORD simply introduces new matter into it.

As to the true nature of matter, nuclear physicists are still struggling to comprehend how energy becomes matter. Their latest hope rests on the newly discovered "Higgs boson" as the subatomic particle that will complete the puzzle (see Robert Lea, "Higgs boson: The 'God Particle' Explained," at https://1ref.us/gj1). Likewise, we still cannot explain the fundamental nature of positive or negative charges, of magnetism, or of gravity. When it comes to human knowledge, we still "see through a glass darkly" (1 Cor. 13:12).

The Bible teaches that the LORD placed our world in a pre-existing universe (Gen. 1:2). Unfortunately, the English translation of the Hebrew word *shamayim* in Genesis 1:1 as "heaven" creates a lot of confusion, as it opens the possibility that the entire universe was created on day one of creation. However, *shamayim* is the Hebrew word for the visible heavens, the sky, or the atmosphere. Bible translations exist in which Genesis 1:1 reads: "In the beginning God created the sky and the earth" (NCV).

It is entirely possible that on day one of creation, the LORD created not only the earth, but the entire solar system with an unignited sun holding eight planets and some 150 moons in orbit. Then, on the fourth day, the LORD ignited the sun, illuminating the earth, the moon, and the planets.

In Isaiah 45:18 we read that the LORD did not create the earth in vain, rather He formed it to be inhabited. Thus the question: Did the LORD create "in vain" the "earth-like" planets—Mercury, Venus, Mars—all of which appear to be barren and lifeless? And what about the large gaseous planets—Jupiter, Saturn, Uranus, and Neptune—surrounded by numerous planet-sized moons. What are they doing out there?

The Creator never does anything "in vain." The earth-like planets and some of the moons of the outer planets of the solar system may be in the first stages of being converted into habitable places for humanity. According to the LORD's original plan, the unfallen earth eventually would have been covered with replicas of the Garden of Delight (*Education*, p. 22). This would have necessitated the expansion of humanity to the neighboring planets. Thus the Creator would have converted these planets into livable places in the full view of onlooking humanity, so that we, too, could have shouted for joy!

> But the formation of millions of living organisms—from the amoeba to the zebra—declares His indescribably inventive genius and compassionate care.

The gaseous compositions of Jupiter and Saturn hint that they may be unignited, miniature suns. The moons of the outer planets would require additional light and energy to support human existence. Therefore, with Jupiter and Saturn becoming small suns, in the unfallen earth, we would have enjoyed not one but perhaps three suns in the sky.

It is inconceivable that on the pre-Flood world anyone looking up could have been blinded by the sun. The Bible states that originally the upper atmosphere contained a water canopy (Gen. 1:7). This must have acted as a disperser of sunlight. Consequently, there were no shadows on the pre-Flood earth.

With the water canopy in place, possibly the light intensity of the uniformly lit blue sky would have varied during the day, starting out slightly pinkish, then turning brighter and brighter blue, reaching its peak at noon. Then, in the afternoon, the process would have reversed, with the blue sky becoming less and less bright and finally turning uniformly pinkish at sunset. During moonless nights, there may have been more stars visible.

On the sixth day of creation, when everything was accomplished, "Then God saw everything that He had made, and indeed it was very good" (Gen. 1:31, NKJV). This was not the first time the LORD created a world and populated it with living beings. There may be untold numbers—perhaps billions—of other created worlds floating in the universe. If the LORD simply repeated what he had done billions of times before, there would not have been a need for this final quality control. That the Creator carefully examined all His creation on earth indicates that we are not like any other created worlds. We are unique!

At the same time, we must be compatible with the rest of the universe and capable of interacting with other created non-humans and even learning their language. Since the universe has been around a long time before our appearance, we will have a lot of catching up to do in the new earth.

The creation of the sun, the earth, and the planets show the Creator's might. But the formation of millions of living organisms—from the amoeba to the zebra—declares His indescribably inventive genius and compassionate care. All organisms were created to fill a designated place in the rich tapestry of the biosphere's ecosystems. Contrary to evolutionary assertions about the "survival of the fittest," no organism can exist alone. "There is nothing, save the selfish heart of man, that lives unto itself. No bird that cleaves the air, no animal that moves upon the ground, but ministers to some other life" (*The Desire of Ages*, p. 20).

Since the middle of the 20th century, biology (the study of life) has been the most important science. Along with the information explosion regarding life processes comes an increased appreciation of the intricate sophistication of living matter.

Much biological knowledge has been garnered from the study of the model organism, *Escherichia coli*. This bacterium is a normal resident of the human colon. Its most important task is to maintain an anaerobic (airless) environment in the colon so that the numerous anaerobic strains of microorganisms can function adequately. It also assists in the digestive process and provides us with the B complex vitamin biotin. Some rogue strains of *E. coli*, which acquired pathogenic plasmids, have given this organism a bad reputation. Working with harmless laboratory strains of *E. coli* for decades, I was drawn into the sophisticated miniature world of the bacterial cell.

E. coli is an automated self-replicating nano factory, with a product inventory that rivals the largest chemical manufacturers of the likes of DuPont Chemical Company. Its single chromosome consists of two circular intertwined DNA molecules, each consisting of 4.6 million nucleotides. The linear sequence of nucleotides is the genetic information of the organism. Inspecting the nucleotide sequence of this chromosome, one sees a seemingly endless run of four nucleotides, abbreviated as A, G, T, and C, similar to what is shown in Figure 5 (below). The printed version of the chromosomal sequence, using a conventional font size, takes up 1,000 pages of single-spaced lines.

E. coli has more than 4,000 different proteins, and the genetic information of the chromosome instructs the cell how and when to synthesize these compounds. The functions of most of the 4,000 proteins are to promote individual chemical reactions (Haitao Zhang and George T. Javor, "Identification of the ubiD gene on the *Escherichia coli* chromosome," *Journal of Bacteriology*, Vol. 182, Iss. 21 [Nov. 2000], pp. 6243–6246).

Among the many substances *E. coli* manufactures is CoenzymeQ or ubiquinone. This substance helps the cell's growth in the presence of air. My laboratory studied the biosynthesis of ubiquinone for about nine years. We discovered the location on the chromosome of one of the genes of biosynthesis of ubiquinone. This gene, called *ubi*D, was at 86.71 minutes on the *E. coli*'s chromosome. (Scientists divide the circular chromosome of *E. coli* evenly into 90 sections, called "minutes.") Figure 5 shows the sequence of the 1491 nucleotides of *ubi*D gene, and, interestingly, shows a change of a single nucleotide inactivated the *ubi*D protein.

These small details are offered here to illustrate the extent of specificity and complexity required for the life processes of this microscopic organism. The totality of information on *E. coli* is now so enormous that it can only be stored in databases such as "EcoCyc" (*Encyclopedia of E coli Genes and Metabolism*, Peter Karp, curator, Internet).

As *E. coli* represents an insignificant fraction of all living organisms, so the vast database on this microorganism is but a drop in an ocean of all biological facts, known and unknown. Our great Creator is the Inventor and Implementer of all biology. The exceptions are all the noxious and toxic entities that are the result of corrupting and parasitizing God's good creation by Satan (see Matt. 13:28). Biology, too, bears the scars of the great controversy.

Let us find reasons for our faith in our great Creator, by being more sensitive to the beauties and wonders of existence! In our innermost being, let us magnify the LORD for His greatness and goodness! When we do so, we are practicing for eternity as this will be the joyous privilege of the redeemed forever.

Figure 5
These 1491 nucleotides make up the *ubi*D gene of *E. coli*. In the third line from the bottom, one nucleotide is underlined. In a mutant strain of *E. coli*, this nucleotide, **G**, is changed to an **A**, which in turn results in an inactive *ubi*D protein. Consequently, the mutant cell cannot utilize certain organic substances for growth (such as succinate), which require air for their metabolism.

ATGGACGCCA TGAAATATAA CGATTTACGC GACTTCTTGA
CGCTGCTTGA ACAGCAGGGT GAGCTAAAAC GTATCACGCT
CCCGGTGGAT CCGCATCTGG AAATCACTGA
AATTGCTGACCGCACTTTGC GTGCCGGTGG GCCTGCGCTG
TTGTTCGAAA ACCCTAAAGG CTACTCAATGCCGGTGCTGT
GCAACCTGTT CGGTACGCCA AAGCGCGTGG CGATGGGCAT
GGGGCAGGAAGATGTTTCGG CGCTGCGTGA AGTTGGTAAA
TTATTGGCGT TTCTGAAAGA GCCGGAGCCGCCAAAAGGTT
TCCGCGACCT GTTTGATAAA CTGCCGCAGT TTAAGCAAGT
ATTGAACATGCCGACAAAGC GGCTGCGTGG TGCGCCCTGC
CAACAAAAAA TCGTCTCTGG CGATGACGTCGATCTCAATC
GCATTCCCAT TATGACCTGC TGGCCGGAAG ATGCCGCGCC
GCTGATTACCTGGGGGCTGA CAGTGACGCG CGGCCCACAT
AAAGAGCGGC AGAATCTGGG CATTTATCGCCAGCAGCTGA
TTGGTAAAAA CAAACTGATT ATGCGCTGGC TGTCGCATCG
CGGCGGCGCGCTGGATTATC AGGAGTGGTG TGCGGCGCAT
CCGGGCGAAC GTTTCCCGGT TTCTGTGGCGCTGGGTGCCG
ATCCCGCCAC GATTCTCGGT GCAGTCACTC CCGTTCCGGA
TACGCTTTCAGAGTATGCGT TGCCGGATT GCTACGTGGC
ACCAAGACCG AAGTGGTGAA GTGTATCTCCAATGATCTTG
AAGTGCCCGC CAGTGCGGAG ATTGTGCTGG AAGGGTATAT
CGAACAAGGCGAAACTGCGC CGGAAGGGCC GTATGGCGAC
CACACCGGTT ACTATAATGA AGTCGATAGTTTCCCGGTAT
TTACCGTGAC GCATATTACC CAGCGTGAAG ATGCGATTTA
CCATTCCACCTATACCGGGC GTCCGCCAGA TGAGCCCGCG
GTGCTGGGTG TCGCACTGAA CGAAGTGTTTGTGCCGATTC
TGCAAAAACA GTTCCCGGAA ATTGTCGATT TTTACCTGCC
GCCGGAAGGCTGCTCTTATC GCCTGGCGGT AGTGACAATC
AAAAAACAGT ACGCCGGACA CGCGAAGCGCGTCATGATGG
GCGTCTGGTC GTTCTTACGC CAGTTTATGT ACACTAAATT
TGTGATCGTTTGCGATGATG ACGTTAACGC ACGCGACTGG
AACGATGTGA TTTGGGCGAT TACCACCCGTATGGACCCGG

CGCGGGATAC TGTTCTGGTA GAAAATACGC CTATTGATTA
TCTGGATTTTGCCTCGCCTG TCTCCGGGCT GGGTTCAAAA
ATGGGGCTGG ATGCCACGAA TAAATGGCCG**G**GGGAAACCC
AGCGTGAATG GGGACGTCCC ATCAAAAAAG ATCCAGATGT
TGTCGCGCATATTGACGCCA TCTGGGATGA ACTGGCTATT
TTTAACAACG GTAAAAGCGC C

CHAPTER 11
It is a Wonderful Life!

Life is a unique property of unique matter. When we take apart the cell, which is the most fundamental level of life, we end up with complex—but utterly inert—lifeless matter. Thus, when we think about life, we are participating in the amazing phenomenon of matter contemplating its own existence!

Two texts in the Bible, taken together, indicate that the earth was made for mankind's sake: "For thus saith the LORD that created the heavens; God himself that formed the earth and made it; he hath established it, he created it not in vain, he formed it to be inhabited" (Isa. 45:18). "The sabbath was made for man" (Mark 2:27). Just like the Sabbath, the world was made for man to inhabit.

We were provided with five ways to sample our world—all in "deluxe" mode—with color vision in three dimensions, stereophonic audio, heat and touch sensors, exquisitely sensitive gas detectors, and a taste monitor. We are fully equipped to appreciate the many gifts the Creator provided for us.

We were also given needs. We experience hunger, thirst, a desire for companionship, and the desire to live forever. The Creator saw to it that all these needs could be fulfilled.

All organisms were created to support each other. "There is nothing, save the selfish heart of man, that lives unto itself. No bird that cleaves the air, no animal that moves upon the ground, but ministers to some other life. There is no leaf of the forest, or lowly blade of grass, but has its own ministry" (*The Desire of Ages*, p. 20).

How contrary this concept is to the current popular evolutionary teaching, which would have us believe that the defining characteristic of life is a struggle for survival. The science of ecology teaches that each organism is vital for the wellbeing of the biosphere.

Looking up to a star-filled night sky, we understand that we are part of a wonderful and gigantic universe. But what are we doing here? We know that the Creator does everything extremely well and never without a very good reason. What is the "added value" to the universe—now that we have made our appearance here?

We were not snuck into the universe. There was, undoubtedly, advanced notice of our creation throughout the worlds. "All heaven took a deep and joyful interest in the creation of the world and of man. Human beings were a new and distinct order" (*Review and Herald*, Feb. 11, 1902, Art. A). "Man was to bear God's image, both in outward resemblance and in character" (*Patriarchs and Prophets*, p. 45). The following observation may be a speculative thought, but, because of our resemblance to the Creator, perhaps unfallen human beings were designated to be representatives of the LORD to the numerous worlds.

The reaction to our creation was that "the morning stars sang together, and all the sons of God shouted for joy" (Job 38:7). Contrary to the thinking of typical Hollywood science fiction, outer space is not populated by menacing aliens. No lesser person than the eminent British theoretical physicist Stephen Hawking stated that his mathematically inclined mind found it a reasonable concept that there were other life forms than humans in outer space. However, we should try to avoid them, warned Hawking, because "if aliens visit us, the outcome would be much as when Columbus landed in America, which didn't turn out well for the Native Americans." A current Discovery Channel series depicts an imagined universe featuring alien life forms in huge spaceships on the hunt for resources after draining their own planet dry. "Such advanced aliens would perhaps become nomads, looking to conquer and colonize whatever planets they can reach."

Nothing could be further from the truth of reality as depicted in the Scriptures. The universe is happy to have us. It is earth, in its fallen state, that is the only blight among the myriads of perfect worlds. Marauding and fighting aliens are unknown entities there.

The Creator is not in the business of mass-producing identical constructs. We may be certain that everything on earth is novel and different from what is found on other planets. So, to get a sense of the magnitude of the Creator's inventive genius, we need to multiply by billions the millions of variations on the theme of life here on earth.

Visitors from elsewhere in the universe would probably enjoy a visit to the local vegetable and fruit markets on earth. It is possible that none of the goods found there exist elsewhere in the universe. When, in Matthew 26:29, Jesus is quoted as saying that He will not drink grape juice until His return, it could be because there may be no grape juice in heaven!

On the other hand, everything on earth is likely to be compatible with beings on other worlds. It would be unthinkable to imagine that we could not communicate with non-humans. There may be a universally understood language throughout the cosmos. How wonderful it would be to know it! What about the music, art, and technology available "out there"? The redeemed have a lot to look forward to in catching up with the rest of the universe.

> *We need to halt our business every once in a while, take stock of all the gifts that we receive from the dear LORD moment by moment; probing this wonderful world with all our sensors, basking in the warmth of human love of family and friends and in the warmth of love of our Great LORD.*

Wonderful as the universe out there may be, we need not sell short our own bailiwick. Every person born on earth is already the winner of a great prize, experiencing being alive with all its potential!

We need to halt our business every once in a while, take stock of all the gifts that we receive from the dear LORD moment by moment; probing this wonderful world with all our sensors, basking in the warmth of human love of family and friends and in the warmth of love of our Great LORD.

Heavenly beings, neither created in the image of God, nor redeemed from eternal extinction by Jesus, live in an attitude of ceaseless praise of God. How much more should we praise Him constantly in our thoughts and live in a spirit of gratitude!

CHAPTER 12

Materialistic or Superintended Creation?

Creationists have their hands full not to be swept away by the gale force winds of evolution blowing through the halls of academia. Undergirding the evolutionary bluster is the firm conviction and battle cry: "Nothing in biology makes sense except in the light of evolution" (Theodosius Dobzhansky). Creationism is dismissed out-of-hand as "anti-science," and creationists are marginalized.

Not surprisingly, creationists expend maximum effort and energy in apologetic works, demonstrating the validity and even superiority of biblical creationism. In my years of teaching biochemistry and microbiology to undergraduates, and to graduate medical and dental students, I integrated creationism into the curriculum, stressing the sophistication and complexity of biochemical and biological systems. At the conclusion of the main biochemistry course for graduate students, I gave two lectures on the impossibility of life arising spontaneously under any conditions and demonstrating that the very existence of life was proof for creation.

For thirty-seven years, I was under the impression that I promoted biblical creationism by my lectures and by occasional articles I published on this subject. But several years after I retired from teaching, it occurred to me that by dwelling only on what the Creator accomplished originally, I implied that the created world and everything in it were exquisitely sophisticated machines.

> *it occurred to me that by dwelling only on what the Creator accomplished originally, I implied that the created world and everything in it were exquisitely sophisticated machines.*

My creationist-oriented message was (unintentionally) that although the LORD can be justly credited for our design and origins, our day-to-day existence is governed by the laws of chemistry and physics. These laws appear to be essentially adequate to explain all physical phenomena. There was no difference between my description and that of a materialist regarding how our world operates in the here and now. Created matter is independent of the Creator. I now call such ideology *materialistic creationism*.

This stance avoids even the appearance of pantheism, the notion that God is in everything. It is supported by everything that science has discovered. But, is it in harmony with the biblical view of existence?

The biblical references to the LORD's involvement in the created world are unambiguous. "The Son radiates God's own glory and expresses the very character of God, and he sustains everything by the mighty power of his command" (Heb. 1:3, NLT). "If God were to take back his spirit and withdraw his breath, all life would cease, and humanity would turn again to dust" (Job 34:14, 15, NLT) and "He himself gives life and breath to everything, and he satisfies every need ... For in him we live and move and exist" (Acts 17:25, 28, NLT).

There are also pointed comments on this subject by Ellen White. The selection from *Testimonies for the Church*, vol. 8, pp. 259, 260 that I quoted at length in chapter 7 is clear. "God is perpetually at work in nature."

Despite this, our current understanding of the interdependence of organisms in the biosphere is more akin to the depiction in Figure 2 (see chapter 7). This is a diagram of a solar energy-driven machine. According to this chart, besides designing and bringing it into existence, there is no further role for the Creator in its operation. Thus, all through my teaching career, I lived with an unresolved tension between the clear statement of the Spirit of Prophecy and my inability to integrate it into my understanding of science.

Following my retirement, I revisited my difficulties regarding the LORD's intimate involvement with the created world in scientific terms. (I mentioned this briefly in chapter 7.) Whereas, until this time I questioned the need for the Creator to push atoms and molecules around to make chemical reactions happen, this time I focused on the very existence of subatomic particles and the forces that control them. Thinking of the fundamental forces—gravity, electrical charges, and the strong and weak forces within the atomic nucleus, which undergird the behavior of matter—I asked, "What if all of these manifestations require the continuous expression of the Creator's power?"

This concept does not advocate that the LORD pushes subatomic particles, atoms, or molecules around to make things happen. Rather, the very existence of matter depends on a continuous input of the Creator's power. This sustaining power is required at the most fundamental levels of existence. I call this *superintended creationism*.

If this is not so, then we have a situation in which the created universe is independent of the Creator. Such a construct is not far from the materialistic view of the universe, which posits that matter alone is sufficient to account for everything in existence.

More recently I found another quotation from Ellen White, where the word "superintendent" is connected with the word "Creator." See what you think of it.

"Those who have a true knowledge of God will not become so infatuated with the laws of matter or the operations of nature as to overlook, or refuse to acknowledge, the continual working of God in nature. Nature is not God, nor was it ever God.... The natural world has, in itself, no power but that which God supplies.... *God is the superintendent*, as well as the Creator, of all things. The Divine Being is engaged in upholding the things He has created.... It is through the immediate agency of God that every tiny seed breaks through the earth, and springs into life" (Ellen G. White, *Selected Messages*, bk. 1, pp. 293, 294, emphasis added).

Superintended creationism affirms:
1. The eternal Godhead Creator existed from forever past—before time, space, and the universe.
2. Every particle in the universe has been created by God.
3. Every particle in the universe is sustained by the Creator moment by moment.
4. Should the Creator withdraw His sustaining power, the universe would cease to exist.
5. On a personal level, the concept of *superintended creationism* assures us that we are constantly under the care of the Creator. Our very existence shows that we are not forgotten!

Superintended creationism removes the mythical aura of omnipotence from inanimate matter. If matter exists only at the pleasure of the Creator, then it will not undergo endless and continuous "evolution" in millions of different directions. The Creator is deliberate in His work. "For he spake, and it was done" (Ps. 33:9). The apparent chaos and less-then-perfect state of our world and our solar system are consequences of the great controversy and show a degradation of the Creator's original design.

On Sabbaths we can worship the Creator not only for creating and redeeming us but also for sustaining us moment by moment. *Superintended creation* cements our relationship with our LORD, as we confess that, without Him, we cannot exist.

CHAPTER 13

Proving Creation

In the year 2000, it was my privilege to write an entire volume of the *Origins* magazine entitled, "Life, an Evidence for Creation" (*Origins*, Vol. 25, No. 1 [1 Jan. 1998]; the whole volume was printed in 2000). The editor of the magazine wrote an introduction to the volume with the title: "Proving God?"

Here is the last paragraph: "These are strong arguments for a Creator. However, they should not be mistaken for absolute proof. God's existence cannot be proved by science. Many intelligent people have chosen to reject arguments such as those presented here. The point of this presentation is not to claim that we have no other choice than to accept the existence of a Creator but to show that we do have the choice to accept His existence. Not only is it reasonable to do so, but, in view of the properties of living organisms, it is the best choice available" (L. James Gibson, "Proving God?" *Origins*, Vol. 25, No. 1 [1 Jan. 1998], pp. 3, 4).

When I read his remarks, I confess to being taken aback. I felt that understanding the basic principle of life coerces the reader to conclude that (1) life cannot come into existence spontaneously under any circumstance anywhere and (2) no created being has the ability to control individual molecules, which capacity would be absolutely necessary for producing living organisms. So, yes, I felt that the monograph actually did prove God's existence, in spite of the editor's introductory remarks.

Now, twenty-two years later, I still find this introduction—and especially the statement that "God's existence cannot be proved by science"—problematic. The editor, issuing this opinion, did not present arguments refuting the thesis of the monograph but stated that "many intelligent people have chosen to reject arguments such as those presented here."

I have not seen any counter arguments. Likely, my editor was thinking of the large majority of science colleagues who automatically reject any reference to the supernatural. My essay mostly described some of the biochemical characteristics of living matter. The conclusion flows from incontestable scientific facts. A "coercive proof" is a thesis that has no viable alternatives.

> *What I considered coercive proof for the origin of life did not appear to coerce my evolutionist or agnostic colleagues, nor did it change the opinions of those who were just not interested in the subject of origins.*

However, what I considered coercive proof for the origin of life did not appear to coerce my evolutionist or agnostic colleagues, nor did it change the opinions of those who were just not interested in the subject of origins.

For many generations, the education and practice of scientists avoided any consideration or reference to the supernatural. This is understandable, on the one hand, because the supernatural is beyond the reach of laboratory experimentation. Yet, on a rock-bottom fundamental level, the ultimate aim of science is the understanding of how things work in nature. When it comes to explaining the origin of life, science ran out of any conceivable naturalistic options, when, in 1865, Louis Pasteur received the Alhumbert prize (2,500 francs) for proving the impossibility of spontaneous generation of life (George T. Javor, "The Scandal of Biochemical Evolution: A fight with Louis Pasteur," *Adventist Review*, July 29, 2022).

Currently the official, academic explanation of the origin of life, found in Wikipedia and taught in secular universities everywhere, is the biochemical evolutionary fairy tale. It is based on seventy-five years of utterly futile laboratory research with no indication that it will ever succeed (see https://1ref.us/gj2). Yet the academic world mindlessly soldiers on, condemned to perpetual failure.

The lack of knowing how life started on earth is only the first hurdle in the quest to explain the origins of the biosphere. Evolutionary theories of the "tree of life" are thoroughly inadequate to account for the complex, multifaceted ecological systems now functioning all around us. The denial of the existence of the all-powerful, wise Creator by "official, academic science" is a prime example of the innate stubbornness of the human psyche, the choice of irrationality over observable facts.

An even more powerful illustration of the unfortunate trait of disastrously faulty human reasoning is the baleful history of the first fifteen hundred years of humanity. During this period, everyone knew their origins. For over eight centuries, Adam, the progenitor of the human race, passed on the authentic story of creation as he heard it from the Creator Himself. Moreover, there was the entrance to the Garden of Eden for anyone to visit and see the angel with the flaming sword, guarding it from anyone's entrance (Gen. 3:24). After the Flood, the survivors retained the seven-day cycle of the week, and, when languages became multiplied, the word "Sabbath," or its equivalent, remained the name for the seventh day of the week in more than one hundred major languages (https://1ref.us/gj3).

In light of the overwhelming evidence of the LORD's magnificent creatorship, genius, goodness, and love for mankind, one cannot help wondering, how could humanity veer so sharply away from any semblance of reverence for God?

Did the antediluvians become intoxicated with their wealth and physical and mental prowess and could not submit to the Creator's lordship over them? Such a baleful attitude toward God is consistent with their foolish attempt to defy God by building a gigantic skyscraper after the Flood.

Thus the early history of humanity is an object lesson that coercive proof of the Creatorship of the LORD does not necessarily produce converts and disciples.

Although the authentic story of our origins was preserved by the Hebrew people and their continual observance of the seventh-day Sabbath is still a persistent reminder of creation, the world at large, in harmony with their disdain for God, chooses to embrace a plethora of fanciful legends to account for our origins.

If we do not know where we came from, then we are equally at loss about where humanity is heading. How appropriate that the first angel, in chapter 14 of Revelation, calls for the worship of the Creator! All approaches to the LORD must begin with acknowledging His magnificent creatorship.

This indisputable fact is the chief cause of resistance by vast numbers of people. If humanity evolved over eons of time, then we are not beholden to a Creator. We are free! We may not be certain how we got here, but here we are! We may do as we please!

The consequences of this "freedom" from any guidance are on display in contemporary society on every level, from individual criminals to nations with rogue activities. Their motto is: "Do unto others before they do it unto you." The utterly destructive nature of such an attitude is beyond argument.

Significantly, our very existence on earth depends on the mutual support by all components of the biosphere. The life sustaining oxygen we breathe is provided by photosynthetic plants. We, in turn, manufacture CO_2 for the vegetation, so that it can grow. Plants also utilize the nitrogen in the air with the assistance of microorganisms that reside in their roots.

It is clear that the notion of "survival of the fittest" is not the "modus operandi" of the biosphere. Would it not be wonderful if humanity would imitate nature in this regard and live according to the God-given principle on every level: "Love thy neighbor as yourself"? For this to happen, our self-destructive human psyche would have to yield to heavenly influences.

CHAPTER 14

The Scandal of Biochemical Evolution

When it comes to the origin of life on earth, there are only two possibilities: either life was created by the Creator or life developed spontaneously from inanimate matter. Until recently there was a third possibility—that life came to earth fortuitously from another source in the cosmos. However, extensive search for life in the solar system showed that only our globe is covered with living organisms and the rest of our cosmic neighborhood is completely sterile. As the closest star system beyond the solar system is 4.37 light-years (more than 25 trillion miles) away, this third possibility has been abandoned.

As for spontaneous generation of life on earth, the French Academy of Sciences awarded the Alhumbert Prize (consisting of 2500 franks) to Louis Pasteur, in 1862, for conclusively showing that it could not happen (John R. Porter, "Louis Pasteur: Achievements and Disappointments, 1861," *Bacteriological Reviews*, Vol. 25, Iss. 4 [Dec. 1961], pp. 389–403).

Thus, there are no meaningful alternatives to the biblical story of creation. Moreover, two indisputable evidences, coming from antiquity, prove that there was a period in human history in which everyone accepted the biblical account of our origins. The first is the almost universal observation of the seven-day cycle of the week throughout most of history.

Modern historians tell us that the weekly cycle originated with the Sumerians and the Babylonians, and that Moses copied the Babylonian calendar when he wrote his books in 500 BC (Wikipedia, "week," Robert Coolman, "Keeping time: origins of the days of the week," 2014).

These assertions are wrong for several reasons. First, we know from the Bible that Solomon began building the Temple in 965 BC (2 Kings 24:13), 480 years after the Exodus (Exodus 12:40). Thus, Moses was alive in 1445 BC, the date of the Exodus!

Secondly, of the 150 to 200 major ancient and modern languages (spoken by more than a million people), more than 100 use the word "Sabbath," or a derivative of it, for the seventh day of the week! (https://1ref.us/gj4).

The only reasonable explanation for this remarkable phenomenon is that there was a period, early in humanity's history, when everyone acknowledged the biblical story of creation.

Amidst the confusion of languages during the building of the Tower of Babel and the dispersal of humanity across the face of the earth, the weekly cycle and the uniqueness of the seventh day were somehow largely preserved.

The singular designation of the seventh day in variations of the Hebrew "Sabbath" in so many languages is not only a resounding validation of the biblical narrative of humanity's origin, but it is also a clear repudiation of any evolutionary narrative.

This curious leftover from antiquity has been ignored by thought leaders of the past one and a half centuries because it does not fit their tiresome "scientific" narrative of our past. Not willing to concede the invalidity of their postulates, they suggest that religious considerations be isolated from scientific facts and placed into a separate "magisterium."

"Scientific narrative" is the code phrase for the postulates and guesses proposed by scientists to find an exception to Pasteur's dictum that spontaneous generation of life cannot happen.

In the 1870s, Charles Darwin suggested that life on earth probably began in a "warm little pond" (Charles Darwin, in a letter to Joseph Hooker, 1871). In the 1920s, John Burdon Haldane and Aleksandr Ivanovich Oparin both proposed that primitive living cells came into existence when organic substances, formed in an airless atmosphere, collected in puddles and interacted with each other (Stéphane Tirard, "J. B. S. Haldane and the origin of life," *Journal of Genetics*, Vol. 96, Iss. 5 [Nov. 2017], pp. 735–739; https://1ref.us/246).

In 1953, ninety-one years after Pasteur's discovery, Stanley Miller, a graduate student in Harold Urey's laboratory at the University of Chicago, published the results of his experiments in which he exposed a mixture of gases (ammonia, hydrogen, methane, and water vapors) to electric discharges and showed the formation of amino acids (Stanley L. Miller, "Production of Amino Acids Under Possible Primitive Earth Conditions," *Science*, Vol. 117, Iss. 3046 [15 May 1953], pp. 528, 529). Because proteins, the all-important substances of living matter, are composed of amino acids, here was a laboratory demonstration of the way life may have originated on earth. Thus the discipline of biochemical evolution was born.

Over the next decades, numerous laboratories jumped into this field and produced an impressive number of biologically relevant substances, using a variety of imagined "primordial" conditions. The one common denominator among these efforts was the exclusion of the gas oxygen because it destroys the desired products.

Practitioners of these biochemical evolutionary efforts completely ignored a surprising discovery by the astronauts of the 1972 Apollo 16 mission to the moon—that vast amounts of oxygen are constantly generated high in the atmosphere by the ultraviolet radiation of the sun acting upon water vapors (News Release, Naval Research Laboratory: 30-72-7; also *Review and Herald*, March 14, 1974). The news release of this discovery suggested that this "photolysis" of water may be the main source of oxygen in the atmosphere instead of photosynthesis. Since this phenomenon has been present throughout earth's history, the concept of an oxygen-free atmosphere any time was rendered null and void.

In 1974, Stanley Miller wrote: "We are confident that the basic process [of chemical evolution] is correct, so confident that it seems inevitable that a similar process has taken place on many other planets in the solar system.... We are sufficiently confident of our ideas about the origin of life that in 1976 a spacecraft will be sent to Mars to land on the surface with the primary purpose of the experiments being a search for living organisms" (Stanley L. Miller, *The Heritage of Copernicus*, ed. Jerzy Neyman, 1974, p. 328).

Indeed, in the summer of 1976, two sophisticated spacecrafts, the "Viking" landers, settled on Martian soil, 4,600 miles apart from each other, and began their search for living organisms. The results stunned the scientific community. Not only were there no living organisms on Mars, but there were no organic substances at all in the soil! These results completely repudiated biochemical evolutionary predictions (Cyril Ponnamperuma, Akira Shimoyama, Masaaki Yamada, Toshiyuki Hobo, and Ramsay Pal, "Possible Surface Reactions on Mars: Implications for Viking Biology Results," *Science*, Vol. 197, Iss. 4302 [29 July 1977], pp. 455–457; Gerald Alan Soffen, "Scientific results of the Viking missions," *Science*, Vol. 194, Iss. 4271 [11 Dec. 1976], pp. 1274–1276).

Near the end of the 20th century, a new field of inquiry emerged—synthetic biology. Here laboratories are attempting to create living organisms! There is widespread optimism that with this development we are at the cusp of a new and exciting era of biology! (Jack W. Szostak, David P. Bartel, Pier Luigi Luisi, "Synthesizing life," *Nature*, vol. 409 [18 Jan. 2001], pp. 387–390).

However, these scientists have bumped into a hitherto largely overlooked aspect of living organisms. The phenomenon of life is based on continuous chemical reactions within each cell, and no one has the technology to construct cells with continuous ongoing chemical processes within! Such a feat requires an ability to simultaneously control myriads of molecules. Only the Creator has such a capacity! (George T. Javor, "Synthesizing Life in the Laboratory: Why is it not Happening?" Geoscience Research Institute Website, July 26, 2021, at https://1ref.us/gj5).

The study of ecology, the relationship between different types of organisms in the biosphere, revealed that no single kind of organism can survive by itself on earth. Plants depend on mammals for CO_2 to make sugar and oxygen through photosynthesis. Bacteria, living in root nodules of plants, convert the nitrogen in the air to nitrate salts, so that plants utilize it for growth. It is overwhelmingly clear now that the various organisms of our biosphere do not compete with each other, but, rather, they form an obligatory network of mutual support.

Thus, if a single organism were to emerge miraculously through putative evolutionary means, it could not survive in the absence of a supportive biosphere!

Since 1862, there has been no new scientific discovery that would invalidate Louis Pasteur's annihilation of the theory of spontaneous generation. The enterprise of biochemical evolution, from 1953 to the present, constitutes a futile effort to find an exception to Louis Pasteur's dictum.

The scandal of biochemical evolution is that, despite its long history of scientific failures and a lack of prospect of ever succeeding, the official scientific establishment still insists on teaching it to students of all ages as the gospel truth! Moreover, through one of its organizations, the American Association for the Advancement of Science (AAAS) it actively crusades against teaching creationism in science classes (George T. Javor, letters to *Microbe*, Vol. 3 [Nov. 5, 2021]; since access to the archives of *Microbe* magazine requires being a member of the American Society of Microbiology, the letter is reproduced below). Teaching this untenable notion, in turn, has deprived generations of students of the knowledge of their true origins and has wasted untold billions of dollars on hopeless projects!

Letter to *Microbe* magazine

Evolution in the Classroom

Risking the ire of the National Academy of Sciences, attention needs to be called to the irony of their current crusade against creationism in science classrooms. Sir Francis Bacon, who is credited with formulating and establishing the scientific method, was a creationist. So were Sir Isaac Newton, Robert Boyle, Louis Pasteur, Carl Linnaeus, Michael Faraday, Blaise Pascal, Lord Kelvin, James Clerk Maxwell, Jean Louis Agassiz, Rudolph Carl Virchow, Johannes Kepler, and numerous other intellectual giants on whose shoulders stand the modern scientific enterprise. Clearly, creationism did not hinder the scientific work of these greats, rather it encouraged them to seek keener insights into the secrets of the physical realm. Permitting students to peek outside the box of evolution is hardly a dilution of science. Rather it is granting them freedom of imagination and thought similar to what students of previous generations were allowed to have.

George T. Javor,
Loma Linda University School of Medicine,
Loma Linda, California.

CHAPTER 15
Consequences of Creationism

Creationism is not for the fainthearted. That educated individuals still cling to a paradigm first proclaimed 3,500 years ago without any proof from some university physics, chemistry, biology, or mathematics department, is, to put it mildly, unorthodox. How much safer it is to take the words of scientists who devote their lives to the careful study of various aspects of nature and who declare in one voice that we are here as the result of a huge explosion of primeval matter many billions of years ago.

Contemporary scientific thinking is characterized by the famous Theodosius Dobzhansky dictum, "Nothing in biology makes sense except in the light of evolution." The introduction to a recent special issue on evolution in *Science* magazine says, "Evolution is the mechanism producing the diversity of life.... Furthermore, the intellectual concepts arising from our understanding of evolution have enriched and changed many other fields of study. Dobzhansky's famous dictum that 'nothing in biology makes sense except in the light of evolution' is even more true today than it was a half a century ago" (Brooks Hanson, Gilbert Chin, Andrew Sugden, and

Elizabeth Culotta, *Science*, Vol. 284, Iss. 5423 [25 June 1999], p. 2105). In the same issue of *Science*, Stephen Jay Gould wrote: "Organic evolution ... [is] one of the firmest facts ever validated by science" (Stephen J. Gould, p. 2087).

The standard creationist response to such declarations is to point out flaws in the evolutionary arguments. But creationists are at their best when they show that their explanations work better than the evolutionary ones. Creationists should develop their paradigm until people conclude, "Nothing in biology makes sense except in the light of creationism."

Some observations about creationism are offered below by way of answering questions about this paradigm.

1. *Is creationism a religiously motivated paradigm?*

Yes. Efforts to present creationism in secular wrapping distort the thrust of creationism. At the very center of creationism is the Creator. A conservative reading of the Bible informs one of an intimate involvement of the Creator with nature and strongly suggests that religion cannot be divorced from science.

Of all the great civilizations, that of Western Europe, with its emphasis on experimentation and mathematical formulations, generated the sort of scientific thinking that we now call modern science (Nancy R. Pearcey and Charles B. Thaxton, The Soul of Science: Christian Faith and National Philosophy, 1994). Through sheer know-how and rule of thumb, several cultures of antiquity— the Chinese and Arab cultures among them—produced higher levels of learning and technology than medieval Europe. Yet, it was in Europe that modern science was born. Heavily contributing to the birth of modern science was the Judeo-Christian faith, with its confidence in the lawfulness of nature.

The supposed conflicts between religion and science are a recent Invention and a distortion of historical realities by a class of historians (led by John William Draper and Andrew Dickson White) who had an agenda to destroy the church's influence (Hanson et al.). The currently popular secularism in science may only be a detour in the history of science.

2. **What are the perceived liabilities of creationism?**
    A. Creationism originated in a prescientific ancient world where many myths persisted. The biblical story of the creation coexists with the Babylonian creation story among others.
    B. Creationism rests on the notion that there is a supernatural being. This assertion cannot be verified scientifically. Moreover, if it is true, then ours is a capricious world, subject to the whims of supernatural powers. Science is not equipped to study such a world.
    C. Creationism restricts the range of inquiries because, by definition, there is no point in studying the origins of life and other interesting questions.
    D. Creationism implies accountability. Humanity is not the supreme authority in the world.

3. **What are the responses to these observations?**
    A. Finding the concept of creation among different ancient cultures certainly suggests a common original source for these stories.
    B. The Supreme Being of the Bible created a world with laws that either were revealed or could be discovered. Humanity is mandated to subdue and care for the creation, using these laws. There appears to be no caprice in the routine operation of nature. Nevertheless, the creationist paradigm permits divine intervention in nature, where known natural laws are superseded. Creationists point out that past divine interventions of great significance were explained to humanity by special revelation. Modern science went astray precisely when it discarded supernaturally revealed information relevant to science.
    C. Whether the creationist paradigm is restrictive has to do with one's perspective. It is true that one's understanding of reality will dictate one's range of inquiry.

4. **Is science hindered or helped by creationism?**
    The creationist worldview was a strong motivating factor to study nature for scientists in the Middle Ages. The thought was to experiment to see

how God ran the world. Those holding this thought were the voluntarists who opposed Aristotelians and the theory that the universe and everything in it had to be made by laws of logic, which Aristotle himself "discovered." Prominent voluntarist scientists who practiced scientific experimentation and measurements were Jan Baptista van Helmont, Robert Boyle, and Isaac Newton.

The biblical doctrine of creation ensured its followers that we live in an orderly world ruled by the Supreme Lawgiver. This is in stark contrast to the pagan worldview, which saw nature as being alive and moved by mysterious forces. Thus, the doctrine of creation was a positive and possibly a decisive contributing factor to the birth of modern science.

5. Is there explanatory power in creationism?

Science, to a great extent, is about explanation. The acid test of the value of a paradigm rests in its explanatory power. Here are some examples:

    A. Elements of design, seen in nature at every level, follow naturally from creationism.

    B. The diversity among organisms is viewed as a reflection of the Creator's generous personality.

    C. Interaction between and mutual support among organisms is in harmony with a benign design.

    D. The burden to explain how living matter came into existence is lifted. So is the burden of having to connect every organism to each other through a phylogenic "tree of life."

    E. This is especially helpful in light of our understanding of the exceptional fidelity of genetic reproduction, on the one hand, and of the very limited range of possible changes that can be accomplished by mutations, on the other. It has now been shown in the laboratory that the bacteria E. coli remains E. coli after thousands of generations.

    F. Not all manifestations of the biosphere have to do with survival values. There apparently is more to life than mere survival. If survival were the only criterion for life, we would see a much starker and sparser world. Creationism also frees us from having to explain why there are both unicellular and multicellular organisms and why there is an absolute requirement for two different genetic types of organisms (male and female) for coexistence.

G. Common features among organisms are understood to come from the same Designer. For example, similarities in metabolic pathways generate metabolic needs, which can be satisfied by common food sources. Diverse features support the ability of organisms to service different niches and to preserve their identities. Differences among organisms also reflect the Designer's obvious penchant for variation.

H. Instead of asking how an organism is successful in carving a niche for itself, we ask, how does this species contribute to the good of the biosphere?

I. The puzzle of the chicken and the egg is solved. The chicken came first.

J. The cause for existence, from atoms up, is understood to be the express will of the Creator. The Creator was not indebted to preexisting matter. Creationists hold that matter is not infinitely old; rather, it had to be created.

K. A characteristic of a designed entity is that the whole is greater than the sum of its parts. Design and organization enable components of complex systems to cooperate for the expression of new functions. In this sense, our entire reality may be shown to have been designed. (See the Figure 1 in chapter 6.)

L. Predation, toxic plants, viruses, suffering, and the death of non-plant organisms do not fit into a scheme conceived by an all-wise Creator. The creationist paradigm assigns these to the work of an evil power in nature. This concept is most helpful when we consider the immense sophistication seen in the operation of living matter, all of which appears to go for naught—that is, to the eventual demise of the organism. It is clear that a truly successful design would permit the existence of living organisms to continue in perpetuity.

6. Can we make scientifically testable predictions using the creationist paradigm?

Creationism has been criticized for not leading to testable predictions. Wrong paradigms may lead to testable suggestions, but that does not necessarily make them good hypotheses. It makes them testable hypotheses.

The Hawaiians thought that the moon can be reached by catching a ride on the moonbeam as it hit the water at moonrise or moonset. So strong rowers would paddle furiously toward the moon at those times. Their lack of success was attributed to not going fast enough.

When a paradigm's prediction is tested and the results are at variance with the predictions, sometimes the paradigm is altered, but more often the test results are reinterpreted to allow for the continuation of the paradigm. When the Viking missions to Mars found no evidence for life on the Martian surface soil even though microbial life was predicted by the chemical evolutionary paradigm, the adjustment was made to postulate the existence of living organisms deep within the Martian soil or the existence of microscopic fossils.

The creationist paradigm suggests that, rather than creating only one or a few species, the Creator generates a rich variety of living organisms. Therefore, it would be surprising to find planets populated with microorganisms alone.

7. What other predictions follow from the creationist's position?
   A. The biosphere is complete. No new orders of organisms are expected to arise. (The creationist paradigm nevertheless is comfortable with new species arising within the same order.) All current organisms have recognizable ancestors.
   B. No living organisms will arise abiotically.
   C. The genetic barriers between "kinds of organisms" cannot be crossed.

8. What are some theological insights from creationism?
   A. Science cannot be divorced from religion. Theologians must not give up the realm of reality completely to the scientist. They may not be able to contribute to the understanding of

how physical realities operate in nature, but they have the grave responsibility of advising scientists on the clearest meaning of supernatural information that has bearing on science.

To illustrate this point, we can imagine a scientist from elsewhere in the universe visiting Earth a week after its creation. Not having been told of the recent creation event and observing the mature organisms and well-developed trees in the Garden of Eden, this well-meaning scientist would justly conclude that Earth had been around for some time.

The conflict regarding the age of the Earth is caused by the fact that dating techniques all but ignore the possibility of a mature Earth appearing suddenly.

- B. Humanity is accountable to the Creator for the way we utilize nature's resources.
- C. The Creator's wisdom and sophistication is documented by countless examples in nature. It needs to be emphasized that He is not only the designer of the world, where objects and organisms are integrated into a coherent setting, but He is the one who brought everything in the world into existence, and He has sustained it for thousands of years. Contrast this to the famous biosphere experiments, which showed the difficulty involved in balancing ecological systems.
- D. Even though we do not have complete understanding of how our world fits into the rest of the universe and what our role in it is, there can be no doubt that the existence of our world has a purpose.

Conclusion

Creationism is a powerful paradigm that is fully capable of undergirding the scientific enterprise of the next millennium. Wider acceptance of creationism by the scientific community in the future may depend on how well theologians can convince scientists as a whole regarding the priceless value of revealed information.

APPENDIX

The Non-equilibrium State of Living Matter Is a Significant Barrier to Laboratory Abiogenesis

Note: I tried to publish this article in a peer-reviewed scientific paper. It was turned down by at least six different journals. The causes for rejection ranged from "lack of space" and "lack of interest" by the readership to just plain rejection with no reason given. I decided to include this article here so that it sees "the light of day" for posterity.

Abstract: Laboratory abiogenesis is one of the ultimate goals of experimental biology. The most formidable barrier to creation of living matter in the laboratory is not the complexity of the cell, but, rather, the absolute requirement for non-equilibrium steady-state for all chemical reactions. Current synthetic biology technologies cannot yet produce cells that harbor chemical systems in non-equilibrium steady-states.

One of the founders of modern biochemistry, Jacques Loeb wrote in 1912: "Nothing indicates, however, at present that the artificial production of living matter is beyond the possibilities of science … we must succeed in producing living matter artificially, or we must find the reasons why this is impossible."[1] This was written prior to the crystallization of a single protein and the discovery of the citric acid cycle and much of metabolism and before understanding the structure of DNA and the concepts of molecular biology. For all these reasons, Dr. Loeb may be excused for his premature optimism regarding the synthesis of living matter. Indeed, more than a century later, the concept of laboratory abiogenesis is just beginning to surface due to the emerging field of synthetic biology.

The ultimate goal of synthetic biology is to produce living matter in the laboratory—in-vitro abiogenesis.[2] Although there is a lack of consensus in the scientific community as to what constitutes "living matter"[3] a very helpful compilation of nine invariant characteristics of all living structures has been published.[4] However, for the purpose of this article, the most helpful characterization of a living entity is given in the insightful review of this field: "… by 'living' we mean the capacity of autonomous self-sustainment in an out-of-equilibrium homeostatic state, with the additional possibility of growth-and-division, giving rise to a sort of minimal life cycle, and to evolution."[5]

Because it is generally thought that at the very heart of all living matter, the "essence of life" is a requisite constellation of nucleic acids and proteins, all current efforts in synthetic biology are directed toward achieving model cells containing these components. Quoting a prominent scientist on this subject: "… life itself can be seen as an emergent property: the molecules that constitute a living cell (DNA, proteins, polysaccharides, lipids, etc.) are not living. The quality of 'life' arises from the assembly of these non-living elements, duly arranged in space and time."[6]

Underlying all efforts in synthetic biology is the fundamentally crucial assumption that it is possible to assemble living matter stepwise from a set of biomolecules. A corollary of this supposition is that, at least in principle, living matter may be disassembled reversibly and reassembled. While such is currently the near universal consensus within the scientific community, these assumptions have not been verified experimentally.

The conviction that living matter may be crafted from inert organic molecules may be traced to the postulates of A. I. Oparin[7] and J. B. S. Haldane[8] regarding the origins of life. With the discoveries of Stanley Miller and Harold Urey,[9] this notion blossomed into the field of chemical evolution, which, over the last three quarters of a century, produced a prodigious body of work[10] but little clarity as to how living matter could have come into existence in the setting of a primitive earth.

Synthetic biology is liberated from the considerable burden of doing biochemistry under primordial conditions. Now laboratories are free to use whatever means they have available to construct living matter! Indeed, there is considerable optimism that synthetic biology will finally accomplish the "holy grail" of biology—the production of an artificial living cell. Accordingly, a concluding remark of a recent review article on the subject states: "The synthesis of a living artificial cell from components will open the door to many more adventurous lines of research.[11]" However, a more cautious reviewer of the evidence states: "... it is important to note that minimal life has not yet been achieved in the laboratory. Does this mean that it is in principle not possible? I do not believe so, although as a scientist it is always good to have a bit of doubt (perhaps we missed something important in our theoretical analysis)."[12]

This communication points to just such an oversight, the underestimation of the essential nature of the "out-of-equilibrium" state of living matter.

All life processes, metabolism, growth, stimulus response and replication are driven by ongoing chemical reactions. Every chemical reaction exists in one of two states, non-equilibrium and equilibrium. Ongoing chemical processes are always in states of non-equilibrium.

When a chemical reaction, aA + bB ⇌ cC + dD runs its course, equilibrium ensues, where the mass action ratio $\Gamma = \frac{[C]^c \cdot [D]^d}{[A]^a \cdot [B]^b}$ becomes the equilibrium constant, K_{eq}. At equilibrium, the change in free energy is $\Delta F=0$ and, in this state, the reaction cannot generate or absorb any energy.

During chemical reactions there is a net flux of matter from reactants to products or the reverse. However, at equilibrium the flux stops.

Moreover, the state of equilibrium resists change. As the Le Chatelier's principle[13] states, if a chemical system at equilibrium experiences a change in concentration, temperature, volume, or partial pressure, then the equilibrium shifts to counteract the imposed change and a new equilibrium is established. Thus, according to this principle, any change from a state of non-equilibrium to equilibrium is irreversible.

Even though, in living cells, each reaction is pushed toward equilibrium by an enzyme (to forestall the possibility of slower, random, non-biological chemical events), if any of the hundreds to thousands of chemical processes could actually reach equilibrium, an irreversible metabolic block would result. Multiple such equilibriums would kill the cell. However, in live cells, there are no isolated reactions, and the problem of equilibrium is avoided. In live cells, chemical events are linked into pathways so that the products of reactions do not accumulate but, rather, immediately react with another substance:

$aA + bB \rightleftharpoons cC + dD \rightleftharpoons eE + fF \rightleftharpoons$ etc

The end products of metabolic pathways are either utilized immediately or they are secreted from the cell. Moreover, regulatory systems such as "feedback inhibition" help maintain homeostasis.

That the non-equilibrium steady-states of all chemical reactions/pathways in live cells constitute an essence of life can be shown to be true by the simple experiment of briefly treating an aliquot of growing *Escherichia coli* culture with drops of toluene.[14] This procedure creates holes in the outer membrane of the bacterium, causing the dissolution of the proton gradient between the cytoplasm and the periplasm.[15] In turn, ATP synthesis halts and within seconds the reactions in the cell reach their equilibriums and the organism dies.

> Current practitioners of synthetic biology, while recognizing the compulsory "out-of-equilibrium homeostatic state" of living matter, do not appear to appreciate the irreversibility and non-spontaneity of the state of equilibrium.

At this point, the dead cell contains most of its nucleic acids, proteins, lipids, polysaccharides, and metabolites. Therefore, while RNA genetic material, enzymes, polysaccharides, and lipids are all necessary components of a live cell, their mere presence is not sufficient for life to exist. In live cells, superimposed on all the necessary biopolymers are the steady-state non-equilibrium dynamics of all chemical events.

Even if it would be possible to prevent internal degradation of its biopolymers, the dead bacterium would never come back to life just by continued incubation.

Current practitioners of synthetic biology, while recognizing the compulsory "out-of-equilibrium homeostatic state" of living matter, do not appear to appreciate the irreversibility and non-spontaneity of the state of equilibrium. Building artificial cells in a modular fashion will inevitably result in the onset of chemical equilibrium within each module. Once equilibrium is reached, the artificial cell, figuratively speaking, "runs into a brick wall." It is no longer capable of growth or accomplishing any net chemical process.

No technology is known to achieve modular assembly of artificial cells while preserving the non-equilibrium status of each component reaction. While these considerations do not apply to polymerizations, such as RNA or DNA synthesis, as each incremental extension of the polymer is accompanied by the hydrolysis of a high-energy bond, rendering these steps essentially irreversible, any other metabolic event is very much subject to termination due to reaching equilibrium. Until the construction of cell-like structures that harbor metabolism in states of homeostatic non-equilibrium becomes a reality, the most sophisticated efforts of synthetic biology will come to naught.

Therefore, more than a century later, our response to Jacques Loeb's call for the synthesis of living matter is that we are not there yet. We need to find ways to generate steady-state non-equilibrium conditions within the artificial cells. Such technologies have not yet been invented.

Endnotes

[1] Jacques Loeb, *The Mechanistic Conception of Life*, 1912, pp. 5, 6; Juli Peretó and Jesús Català, "The Renaissance of Synthetic Biology," *Biological Theory*, Vol. 2, Iss. 2 (20 March 2007), pp. 128–130.

[2] Pier Luigi Luisi, "Toward the engineering of minimal living cells," *The Anatomical Record*, Vol. 268, Iss. 3 (9 Oct. 2002), pp. 208–214; Jack W. Szostak; David P. Bartel; Pier Luigi Luisi, "Synthesizing life," *Nature*, Vol. 409 (18 Jan. 2001), pp. 387–390; Pasquale Stano, "Is Research on 'Synthetic Cells' Moving to the Next Level?" *Life* (Basel), Vol. 9, Iss. 1 (26 Dec. 2019), p. 3.

[3] Erwin Schrödinger, *What is Life?* 1945; Gail Raney Fleischaker, "Origin of Life: An operational definition," *Origins of Life and Evolution of Biospheres*, Vol. 20, Iss. 2 (March 1990), pp. 127–137; Martino Rizzotti, et al, eds., *Defining Life: The Central Problem in Theoretical Biology*, 1996; Kepa Ruiz-Mirazo, Juli Peretó, and Alvaro Moreno, "A universal definition of life: autonomy and open-ended evolution," *Origins of Life and Evolution of Biosphere*, Vol. 34, Iss. 4 (June 2004), pp. 323–346; Gilles Bruylants, Kristin Bartik, and Jacques Reisse, "Is it useful to have a clear-cut definition of life? On the use of fuzzy logic in prebiotic chemistry," *Origins of Life and Evolution of Biosphere*, Vol. 40, Iss. 2 (April 2010), pp. 137–143.

[4] David L. Abel, "Is Life Unique?" *Life* (Basel), Vol. 2, Iss. 1 (March 2010), pp. 106–134.

[5] Stano, p. 3.

[6] Luisi, pp. 208–214.

[7] Aleksandr Ivanovich Oparin, *The Origin of Life*, translated by Sergius Morgulis, 1938.

[8] John B. S. Haldane, *The Origin of Life. The Rationalist Annual for the Year*, 1929.

9   Stanley L. Miller, "A Production of Amino Acids under Possible Primitive Earth Conditions," *Science*, Vol. 117, Iss. 3046 (15 May 1953), pp. 528, 529.
10 Juli Peretó, "Controversies on the origin of life," *International Microbiology*, Vol. 8, Iss. 1 (March 2005), pp. 23–31.
11 J. Craig Blain and Jack W. Szostak, "Progress toward synthetic cells," *Annual Review of Biochemistry*, Vol. 83 (June 2014), pp. 615–640.
12 Pier Luigi Luisi, "Toward the engineering of minimal living cells," *The Anatomical Record*, Vol. 268, Iss. 3 (9 Oct. 2002), pp. 208–214.
13 Henri Louis Le Chatelier and Octave Boudouard, "Limits of Flammability of Gaseous Mixtures," *Bulletin de la Société Chimique de France (Paris)*, Vol. 19 (1898), pp. 483–488.
14 This procedure is the first step in the assay for β-galactosidase. See for example: George T. Javor, Ann Ryan, Ernest Borek, "Studies of the impaired inducibility in relaxed mutants of Escherichia coli," *Biochemical and Biophysical Acta*, Vol. 190, Iss. 2 (22 Oct. 1969), pp. 442–453.
15 Robert W. Jackson and John A. DeMoss, "Effects of Toluene on *Escherichia coli*," *Journal of Bacteriology*, Vol. 90, Iss. 5 (Nov. 1965), pp. 1420–1425; Simon Halegoua, Akikazu Hirashima, Jun Sekizawa, Masayori Inouye, "Protein Synthesis in Toluene-Treated *Escherichia coli*," *European Journal of Biochemistry*, Vol. 69 (1 Oct. 1976), pp. 163–167.

Other Resources
by George T. Javor

Articles

"A Creationist's View of the Solar System" at https://1ref.us/gj6
"Creation in Focus" (aka "Heart and Soul: Theology") at https://1ref.us/gj7
"Creationism: Still valid in the new millennium?" at https://1ref.us/gj8
"Life, an Evidence for Creation" at https://1ref.us/gj9
"Materialistic or Superintended Creation?" at https://1ref.us/gj10
"Teaching Biology in the Light of Creation" at https://1ref.us/gj11
"The Bible and Microbiology" at https://1ref.us/gj12
"The Mystery of Life" at https://1ref.us/gj13
"The Scandal of Biochemical Evolution" at https://1ref.us/247

Books

A Scientist Celebrates Creation, TEACH Services, 2012.
    Purchase: https://1ref.us/gjbk

Evidences for Creation, Review & Herald Publ. Assn., 2005.
    Purchase used: https://1ref.us/gjbk
    View online: https://1ref.us/gjec
The Best News Possible: You May Live Forever! TEACH Services, 2020.
    Purchase: https://1ref.us/gjbk

Video

"The Most Astounding Fact About the Universe" at https://1ref.us/gj14

TEACH Services, Inc.
PUBLISHING

We invite you to view the complete
selection of titles we publish at:
www.TEACHServices.com

We encourage you to write us
with your thoughts about this,
or any other book we publish at:
info@TEACHServices.com

TEACH Services' titles may be purchased in
bulk quantities for educational, fund-raising,
business, or promotional use.
bulksales@TEACHServices.com

Finally, if you are interested in seeing
your own book in print, please contact us at:
publishing@TEACHServices.com

We are happy to review your manuscript at no charge.

www.ingramcontent.com/pod-product-compliance
Lightning Source LLC
Chambersburg PA
CBHW070543170426
43200CB00011B/2531